D0412230

Running Your Own

SMALLHOLDING

Running Your Own

SMALLHOLDING

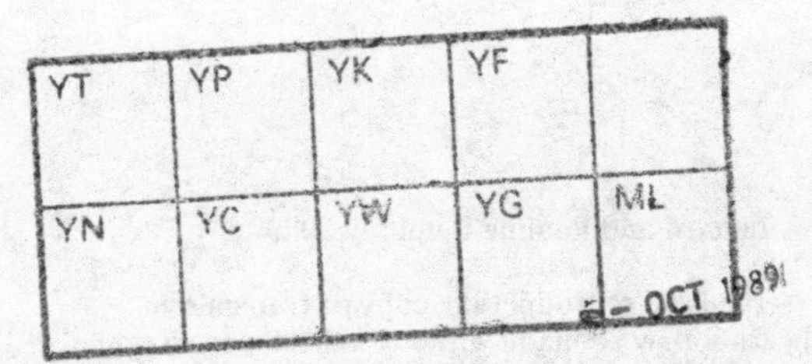

Richard and Pauline Bambrey

KOGAN PAGE

For Garry and Sue, without whose help and advice we would not have got this far.

First published in Great Britain in 1989 by
Kogan Page Limited, 120 Pentonville Road,
London N1 9JN

British Library Cataloguing in Publication Data
Bambrey, Richard
Running your own smallholding
1. Agricultural industries. Smallholdings.
Agriculture - Manuals
I. Title II. Bambrey, Pauline
631

ISBN 1-85091-851-1

Typeset by DP Photosetting, Aylesbury, Bucks
Printed and bound in Great Britain by
Biddles Limited, Guildford

Contents

Introduction

The last one hundred years have seen a more profound change in our way of life than any similar period in history. Many of the changes have been advantageous, but it seems strange that most of our grandparents found their livelihoods in association with agriculture. It is probably this relatively recent bond with nature that makes so many people enthusiastic gardeners. It is the same bond that makes the idea of earning your living by some agricultural or horticultural means so appealing - something small enough to be easily managed, but large enough to provide some kind of a living: in other words a smallholding.

A smallholding is defined as a piece of land used for agricultural purposes, smaller than a farm, typically between 1 and 50 acres. At present, with the large differentials in property prices throughout the UK, it is perfectly possible for someone living in an affluent area to sell a semi-detached house and medium-sized garden and buy a reasonable smallholding. The problem is that buying a smallholding does not make you a smallholder; it is making a living out of the holding that does that. Smallholding is an ancient way of life and it has never been an easy life. But since the Enclosures Act it has become impossible for the smallholder to emulate the larger farmer. The secret of successful smallholding today is the undertaking of small specialised projects, and not trying to earn a living from agricultural means alone.

Chapter 1

Why Become a Smallholder?

People give various reasons for wanting to become smallholders. In general these can be divided into two groups: bad reasons and good reasons. It is well worth examining some of the more common reasons and asking yourself where you fit in.

Some bad reasons

Pressure of business

If you want to become a smallholder to escape the pressures of business, it will not work (without a private income). Your smallholding is going to be your business and to make it pay is going to be at least as hard as any other business. It will probably be far more enjoyable along the way, but do not expect it to be easy. There are lots of other people already doing the same thing; you have to find your own niche in the market.

Dislike of work

This is one that should speak for itself, but to some it obviously does not. Smallholding by its very nature tends to involve many small projects, each with its own chores which mostly have to be done regularly on a daily or twice daily basis. That gives a seven-day week straight away. Many small chores laid end to end with their associated tidying up add up to a 12-hour day. If you do not relish the idea of being tied to the holding seven days a week, 52 weeks of the year, think carefully.

Leading a simple life

This falls into the same category as 'pressure of business'. If you have a private income, life can be as simple as you like. However, if like most people you have to pay the bills from what can be made on the holding, the simplicity is soon replaced by planning, pricing and costing, followed by marketing your produce and estimating safe profit margins while trying to stay competitive.

The truth is that smallholding has never been simple. The rose-tinted ideas created by the media during the 1970s have done great damage to many people who believed it was really that simple. Society as it is at the moment requires individuals to have money available far above and beyond that required to keep body and soul together. This should be borne in mind at all times when considering the financial basis for any new or existing project.

Some good reasons

Working for yourself

Smallholding brings you as close as possible to working for yourself. Every minute of the day (and night) is at your disposal. It is your choice what you do and when you do it; the only major outside influence is what the market wants. You decide what to buy and what to sell, when and where to do it; which job has the highest priority, whether you stop for dinner or continue until darkness falls. In fact every decision that has to be made is yours, from the smallest daily choice to the largest policy decision that could change the direction of your life completely. This has to be one of the most exciting aspects of becoming a smallholder.

Getting down to basics

Smallholding takes you very near the basic requirements for life - food, water and shelter. If it were not for the 'overheads' imposed by society, life could still be that simple. However, it is possible to reduce your food bill by some sensible self-sufficiency. By cutting out the middleman you can produce cheap and wholesome food. Any excess should either be frozen (or preserved by other means) and, if there is an absolute glut, sold. There are always people glad of fresh food - holidaymakers in particular. Because you will be doing what you want to, money spent on leisure activities, ie holidays in the country, will also be reduced.

Having a better lifestyle

This is the reason most often given for wanting to become a smallholder; it is also one of the best reasons for doing so. This 'better lifestyle' doesn't mean peace of mind or money in the bank. What it does mean is freedom of choice, coming in wet and tired with the knowledge of a job finished because you wanted to finish it; watching your holding take shape around you, something new arriving, something old going; walking around your field(s), however big or small, inspecting your stock and making plans for the future; being part of the community (not something that happens at once) and adding something to rural life; and, of course, escaping from the crowds and concrete jungles of the larger towns and cities. Children brought up on a smallholding tend to become far more self-reliant. They start to take on more responsibility and before long they can be trusted to take over a portion of the daily chores.

Those are some reasons for becoming a smallholder. If you find yourself giving some of the bad reasons, it doesn't really matter. What does matter is that you recognise the fact and act accordingly. If you dislike work for its own sake, for example, do not worry. When smallholding there is always a good, obvious reason for the work you are doing.

Aspirations

It is important that you know why you want a smallholding because, without that clearly fixed in your mind, it will be difficult for you to formulate more detailed plans later on. In fact, it is important for the whole family to decide what they hope to gain from the smallholding and to discuss it. This may bring to light some vital point of conflict which can easily be rectified at this stage, but which could cause problems if left until after the holding had been purchased.

Most people who want a smallholding imagine that they will be able to obtain full-time employment and a full income from a few acres, a barn and their own ability. Certainly every hour of the day (and night) can be filled with work without any difficulty. The problem lies with the income. The trend for farms to grow larger and more specialised has been brought about by the search for efficiency and hence profit. It is very difficult for the smallholder to compete in the conventional farmer's market simply because of the difference in scale.

If the possibility of making a full paying agricultural occupation out of your smallholding is slim, you may well ask 'what is the point?' The answer is not to compete in the farmer's market, at least not to the exclusion of all else.

Try also to identify any shortcomings that you may have and plan accordingly. Be honest at this point; it is no good rearing animals for slaughter if you cannot bring yourself to take them to the abattoir, or expect someone to look after chickens if he or she has a phobia about feathers. These may seem rather extreme cases, but there are less obvious ones that are better discovered earlier rather than later.

Individual skills

Try to identify any skills or special abilities which could be of use once you are set up in your holding, but do not bank on being the only person in the district with that skill. Do not merely consider things agricultural because smallholders have traditionally relied on non-agricultural craft work to supplement their earnings. Crafts today should be taken to cover anything from weaving to welding and crochet to computers; in fact, anything that can be run from home, by mail order if possible, that will add to the total income.

Personal background

Smallholders seem to be drawn from a greater diversity of backgrounds than any other group of people. There does not appear to be any general rule concerning success; people with experience in farming can have problems as easily as anyone. Equally, a dedicated novice can run a very successful smallholding. The only general rule is that it will be hard work. This is especially true for someone coming from a desk job, in which case it would be a good idea to prepare physically as well as mentally. It is no good knowing what you want to do and not having the stamina to do it. One thing to note here is that you should not over-estimate your physical ability. It is better to ask for help with one job than be unable to do anything for a week because of a strained back. The other is that although you may struggle at first, it is surprising how quickly you become stronger and need fewer cups of tea (always a good excuse for a sit down).

It is worth looking at your background and trying to identify any areas of experience you have which could be used either directly on the smallholding or indirectly to earn extra money by jobbing.

Smallholding past, present and future

Smallholders or cottagers of the past are often depicted as uniquely happy and contented individuals peacefully going about their daily chores, milking the house cow, tending the bees, feeding the pig, usually with the sun shining – in fact, living an idyllic life. Without a doubt this element is present in smallholding today, but necessarily modified by the demands of modern life.

Our smallholding forebears used to support themselves with

just three acres of land, we are told, but we must not forget the greater availability of common grazing and the simpler lives that these people led. Over the years the common land has been enclosed and improved giving an overall increase in productivity, but to fewer people. Also, society has become more complex and few people today would tolerate the hardships accepted as everyday life in the past; for example, we now have indoor toilets (water rates), electric lights (electricity bills), health care (National Insurance contributions), and so on. In other words, now we have not only to feed and clothe our families and provide a roof over our heads, but we also have to finance a range of overheads. These can be trimmed to a minimum, but this can make life rather dull and cause friction within the family, especially if there are children.

So the first important point is that we cannot model our smallholding on a nineteenth-century one and expect to have a late twentieth-century lifestyle. To obtain this level of affluence from a smallholding today it is necessary to undertake more specialised projects or 'fringe' farming. This often involves the production of high-quality, high-cost products such as quail or crayfish for the gourmet market. At one time mushroom production fell into this category, but now it is another multi-million pound business and of little use to the small producer. The lesson here is that the market is not static, what is a viable proposition one year (month) has been undercut the next.

As more people take up these fringe enterprises and succeed, it becomes progressively more difficult for other smallholders to copy them unless they can find something new and relatively unexploited. There are always new requirements to be fulfilled, but we believe the future of smallholding lies in diversification away from agriculture/horticulture and into other fields such as crafts, tourism/leisure or light engineering.

Cash flow

When planning your smallholding venture it is essential that cash flow is given serious consideration. It is no good ploughing all your working capital into something which will take two years to show a profit if you need a return in six months to stay solvent. The longer-term investments are often more profitable, but only if the bank is not taking interest all the time. The best bet is a balance between quick turnover giving some profit and longer-term projects giving a potentially better return.

It is also essential to cost out any prospective project as

accurately as you can. It is acceptable to say, 'I'll get a piglet and rear it on garden scraps and a bit of barley meal and we can have cheap pork' because the chances are you will get cheap pork and if you don't it is not a crushing blow to your income; in other words it would be a hobby.

If the loss was 10p per pound and the pig killed out at 100 pounds the loss would be £10 - not unbearable. Consider three sows having two litters per year and averaging 10 surviving piglets per litter. That is 60 porkers at 100 pounds losing 10p per pound, totalling a loss of £600 per year. Suddenly a small loss can hurt - three sows is not an unreasonable number and the loss could be *more* than 10p per pound.

It is essential that you know the market you are producing for, how it may change throughout the season and what the overall long-term prospects are.

Self-sufficiency

Self-sufficiency was the dream of the 1970s. Many serious books were written on the subject along with some not so serious ones, and it later appeared in various guises on television. It was mostly treated in a light-hearted manner on television and shown as a great way to escape from the system or to 'drop out', but little useful information was to be gained. In fact, if the programmes were taken seriously, instead of being just good light entertainment, they could be accused of doing harm by over-simplifying the whole issue.

However, taking a more positive view, self-sufficiency has an important role to play in modern smallholding. If you want to categorise self-sufficiency, it is any work that minimises the expenditure of real money; but even in this guise it should be treated with respect. It is easy to become involved with some project or another which should save vast sums of money by generating electricity for free, but this is usually done for entertainment rather than for economic gain. The amount of return is often too small to be useful or the outlay on time and money is not justified by the return. If you can prove that it will pay its way, do it.

There is little doubt that you will be told it is not worth spending an hour making something which can be bought for £1. This is only true if the shop-bought item is as least as good as yours and if you would have earned more than £1 plus deductions during that hour. This applies to repairs of machinery, out-buildings and, of course, your dwelling. There is more money to be saved by doing your own maintenance than by building a methane production plant to cook by - though if you were heating a glasshouse to produce early crops then that could be useful.

However, part of the smallholding mentality is to produce all that you can for yourself, if it can be done economically. Many people (ourselves included), when working out how much hard cash will be required once they have their smallholding, make one serious error. When listing the requirements of life they tend to mark zero against fresh vegetables and meat, making the assumption that what they produce themselves is free. This could not be further from the truth, even if they do not price their time. Vegetables are good value if you do not count your labour costs and if you have a good season; if not, you have to buy seed and vegetables as well. If you enjoy gardening and can afford the time, go about it in a businesslike way and grow exactly what you need and try for production when the crop is expensive in the shops - early potatoes or lettuce, for example.

Meat production cannot be marked down as zero cost because of the feed bills. Whether you are fattening some cockerels or rearing rabbits there has to be a food input. If you scrimp on this, they just take longer to fatten. It is better to allow yourself a 25 per cent saving when doing calculations than assume 100 per cent and be disappointed.

Some people go as far as writing off all dairy products as well, assuming that a cow will provide milk, cheese, butter and yoghurt. She will, if properly fed and you can afford the time to process the milk.

The real problem is knowing when to stop with self-sufficiency. The saving of a few pounds can be such fun and so totally absorbing that it can soon expand into a full-time occupation, which completely halts the less interesting but more profitable pursuits necessary to bring in the hard cash required to live.

Research

The time for research is before you make any positive move such as resigning, selling your house or whatever. The most obvious type of research is to read books. Try to concentrate on factual books dealing with specific areas of interest. Avoid those which glorify the idea of living in a semi-derelict shell, half way up a mountain with a stream running through the house. Build up a file of information, addresses, references and the like. Once running your holding you will no longer have the time for such 'academic' pursuits. If you intend keeping livestock, try to go on a farm holiday where you can actually get your hands dirty. It is much easier to learn about farrowing a pig by being there and actually taking a piglet as it is born and breaking the cord than just reading about it in a book. It is worth visiting markets to see how they function and to follow the bidding as well as getting up-to-the-minute information on prices.

If you feel that you are going to need reference books, buy them before you move because the money will be less available afterwards. For your non-agricultural ventures(s) you need to be fully informed before you move so that the least possible time is lost once you have settled in.

Any specialist venture that you consider undertaking should be extra carefully researched as they can sometimes appear superficially very profitable, because of the lack of information available. Then on further investigation a flaw may be found, throwing all your calculations out. This could be high initial investment, high mortality/failure rate or just a very unpredictable demand for the produce. It is important that you know your venture intimately from set-up to market. Look for other people who are already doing similar things; they may be willing to share their knowledge with you. If they are not, the worst you can get is a blunt refusal and that will not hurt you financially, whereas a serious misjudgement on the saleability of a product could.

Starting

Once all the talking has been done and doubts have been overcome, you have to make a decision to start. It is best to set yourself a timetable. A date has to be set on which you will contact estate agents in one or more areas, another on which you will visit an area, and so on. The only date that should not be fixed is one by which you should have moved because, until you find the right property, it is impossible to tell when that will be. Finding the right property is the easy part; you just have to keep on looking. The difficult part is making the serious decision to embark on smallholding as a way of life.

Finally, going into smallholding with a rose-tinted misconception of it is disaster. Above all, be realistic, make your plans carefully and stick by them unless given 100 per cent proof that they are flawed.

Chapter 2

Planning and Financing Your Smallholding

Choice of enterprise

Once the decision has been made that smallholding is for you, the next step is to formulate some definite plans about what you are going to do on the holding. This choice is completely personal, but must be influenced by market forces. It is useless producing unwanted goods or goods that are produced far more economically elsewhere. The choice of enterprise and the choice of location are inter-related. Having a holding in Orkney and planning to grow early spring flowers will not work out either. Of course, here the cost also plays a large part. As a general rule of thumb the better the climate the more expensive the land, but this is affected by other things like the accessibility of the area from a large conurbation. So you end up with three main considerations: what you would like to do, where it is best to do it and where you can afford to do it. There needs to be some horse trading done here, but do not decide to do something you dislike just for the sake of expediency, because you will have to do it every day.

Time spent in the library is a good investment if only to open up new lines of interest. Browsing through the sections on livestock, natural history and rural interest will usually bring something new to the surface. People have made a living from the most unlikely ventures by identifying a need and supplying it at the right price, which can be surprisingly high for specialist items. Therefore, when you are planning your enterprise, try to select things that are slightly out of the ordinary.

Agricultural v non-agricultural

There is a danger when looking in from the outside to think, 'Well, I just put some calves in the field, they eat the grass and when they get bigger I sell them for a profit. Easy money!' Yes? No, I am afraid it does not work like that. There are the vet's bills for dehorning and castration, injections if they fall sick, milk replacer if bucket reared, cereal until they are old enough to turn out and, when you have successfully reared them, you find that because the export trade for veal keeps the cost of young calves artificially high they are not worth much more than you initially paid for them. Don't be put off; this is only one case, but at the moment it is true. The market fluctuates and the only way to know what is happening is to follow the market reports in the newspapers or on radio and television. It is important to understand that it is no use trying to beat the large farmer at his own game. He can buy food cheaper in bulk and can buy and sell in larger numbers, getting better deals as he does so. Try to identify a gap in the market or possibly a small-output, high-profit area supplying a specialised product. This could be anything from home smoked bacon to snail farming.

There is always a seasonal local trade for Christmas fare such as turkey and goose which is well worth following – also duckling at Easter – but these should be looked on as bonuses as they are so short-lived; your business should be based on less ephemeral requirements. Do not fall into the trap of only looking to the local market. The modern transport network means that with careful planning you can have a national or even international market for your produce.

When you have a definite idea of what you want to do, test the water with a few letters and telephone calls to possible customers, eg hotels and restaurants if you are planning production of table ducklings. You will also need to know how much (and what) a duckling eats to get it to table weight and hence calculate the cost of production.

You will almost certainly find yourself busier during some seasons of the year than others. This is when you should consider the non-agricultural side of smallholding. Smallholders through the ages have supplemented their agricultural earnings by other income (or possibly they supplemented their income by producing food and selling any excess). Traditionally, it was by what is now termed 'rural crafts' using the raw materials found in the countryside and fashioning items to be used locally. Today this could mean making walking sticks if you have a good stand of hazel or possibly restoring old cars if you have the aptitude and an empty barn. These ideas may not fit in with the 'good life' idea of smallholding, but they are one of the keys to success. With the right idea and the willingness to apply yourself to the task in hand there is potentially more money to be made in a 20 ft square workshop than on a smallholding of maybe 20 acres farmed conventionally. When you enter the rural community the diversity of enterprises you find being pursued in barns and old sheds is surprising. There are those that fit in well with the general ideal of smallholdings: pottery, carpentry, jewellery and other similar pursuits. In other barns you may find racing cars or electronic goods supplies and maintenance. The important similarity is that to be successful the business has to be conducted by mail order or through retail outlets, otherwise the tendency is to fulfil the local need quickly and then to rely on passing trade, possibly holidaymakers or newcomers to the district. The exception to this rule is the business that is so specialist that it draws customers from all over the country - or even the world. The racing car example is of this type; although it is a strange companion to smallholding it is a good one because there are few, if any, market forces which affect both simultaneously. It boils down to the old adage 'don't put all your eggs in one basket'.

Integration within projects

Although we have suggested that diversity is one of the keys to success, this must not be taken to mean that your resources are to be spread thinly over a large number of disjointed and totally unrelated projects. You should have a plan with major objectives outlined. Within each area there should be as little waste as possible. For example, if goats' milk production was one of the major aims, rearing pigs could prove to be a useful sideline using up any surplus or spoilt milk. Pork is at present fetching a very poor price, but if you can reduce the cost of production by substituting cheaper feed for the expensive bagged type, you can start to see some profit.

Another example of integration would be the production of fleeces for spinning if fat lamb production was the main objective. It would not be difficult to start in a small way by including a few ewes giving good quality wool in the flock of draft ewes. When shorn these fleeces could be kept back and sold privately or if preferred could be spun, dyed and knitted into the finished garment, The point to beware of is spending an excessive amount of time in producing an article which does not give a proportional financial return. If, however, you have or can find a contact who is interested in marketing this product for you at a good price, it is worth trying.

If you plan to have any kind of livestock you will have dung and bedding as waste; these are the raw materials for worm farming. This admittedly does not sound very promising at first, but has many possibilities. Freshwater angling is a well-supported sport and not all fishermen have their own garden. Worms can be sent by post giving you a large market to work with.

So in short, when planning what will be done on your smallholding, try to see any way the various projects can be linked so that the by-products or rejects from one can be used by another. The only warning is that the second project should make use of the first, but not rely on it totally so that failure of one means failure of both.

Grants and subsidies

Typically, the smallholder does not fare very well from agricultural grants because these are often geared to meet the needs of today's larger farms. Until recently the Agricultural Improvement Scheme (AIS) was the main source of grant money for improving the farm (as its title implies). This allowed percentage grants for various improvements to such things as fencing, and laid down three main prerequisites for anyone wishing to apply; these are that they must:

Earn at least half of his/her annual income from the agricultural business;
Spend at least 1100 hours per year working on the holding;
Either have been farming for at least five years or hold a suitable training certificate. (From form AIS 1(W))

The first two conditions may seem reasonably easy to deal with until one realises how the 1100 hours are estimated. There is a table of standard labour data which details most enterprises undertaken in agriculture and horticulture and how many manhours per year would typically be spent per unit, ie how many hours per year one person would spend on one sheep. The problem here is that the larger unit tends to be more efficient and as this is data obtained in the farmer's domain it leaves the smallholder wondering where all his time is going. The following are some typical extracts from this list.

Item	Hours	Unit
Dairy cows	55	per cow
Ewes	4	per ewe
Laying hens	32	per 100 hens
Deer-breeding hinds	4	per hind
Rough grazing	2	per hectare
Potatoes, first early	200	per hectare
Onions, salad	1,750	per hectare
Rhubarb, forced	37,000	per hectare

Rhubarb, natural	4,400	per hectare
Raspberries	2,000	per hectare
Rabbits	7	per breeding doe

So from the figures quoted above it appears that over 2000 breeding rabbits would take the same time as one acre of forced rhubarb. The AIS scheme has recently stopped operating and is soon to be replaced by a grant system designed to encourage conservation. This will probably still be bound by the same eligibility requirements.

The most useful grant available is the Enterprise Allowance Scheme (see also page 79). This gives £40 per week for one year and also free consultation with business advisers and accountants. The requirements for acceptance on to the scheme are that you have been registered unemployed for at least eight weeks, are in receipt of benefit and can demonstrate that you have £1000 available to put into the business. The requirement of £1000 does not mean that you must have it available in used notes - a bank letter agreeing an overdraft facility to that value is satisfactory. If you have done your homework and can give a confident and logical description of your planned enterprise, the bank manager will be pleased to offer you the overdraft facility because it will bring £2080 (£40 × 52) into your account over the year. You can improve your chances of getting the bank to agree by insinuating that you will not have to use the overdraft anyway and that it is just to bring in an *extra* £2000. If you use the overdraft to buy some piece(s) of equipment, it would be wise to repay the money during the first year as you do not want to enter your first year of non-grant-aided trading with your business account in the red. Not only is this bad business, but it is also very disheartening.

Now that farming is going through a difficult period there is a great deal of interest in diversification. This can embrace anything from holiday bed and breakfast, through traditional country baking to small engineering workshops. In some parts of the country there is a strong interest in encouraging diversification in rural areas. It is worth approaching the local councils in your chosen area(s) to gain information about grants/subsidies available for the type of thing you may have in mind. Usually the type of enterprise which is best received will fill a gap in the area, especially if it has a sound manufacturing basis. There tend to be plenty of second-hand car dealers and mobile mechanics available, and although these would probably succeed in their own right if conducted well, they are unlikely to attract government grant money, mainly because they are not novel and are unlikely to reduce unemployment. At present the most likely way to attract small business grants is by reducing unemployment and bringing some kind of high technology and/or manufacturing to the area. This is not easy to do but it is worth considering if you have the ability.

Finally, a cautionary word on grants. When planning your enterprise, whether or not it is agricultural, it must not rely on grants for success; it must be viable in its own right, otherwise you may fall into the trap of thinking everything is all right until the grant money stops. Then there is only one way your business will go and that is down. However, if the grant is regarded as icing on the cake or insurance against a slow start-up or even a lull in sales after an initial rush, you will not go far wrong.

To obtain any grant you will need to provide certain basic information:

1. You must show that you have good general knowledge of what you are talking about; for example, how to produce an item and how to market it.
2. You must be able to prove that it is a viable proposition; production cost and overheads must be covered by the wholesale price and allow a good profit margin with a realistic mark-up in the retail price.
3. You must produce a financial plan, showing that the business will survive and so will you.

Producing the financial plan

The last section made the smallholding sound very much like a business and that is exactly what it is. The chances are that you will be selling your home and using most or all of the resulting money to buy your smallholding. By so doing you must understand that you are putting the family home on the line. If you fail to make a go of it you must still find income from somewhere and well-paid jobs are scarce in rural areas. Remember this when you make your financial plan and do not delude yourself. Do not try to squeeze a paper profit from something that will not give one in reality. In some circumstances it may be acceptable to add a little rose-coloured polish to the profit already calculated, but never turn a loss into a profit by choosing the figures which give the best results.

Typical of this are the figures quoted for the number of animals per acre of land. It is quite possible to find figures of seven ewes to the acre from some of the agricultural colleges around the country. Remember the flockmasters in charge of those college flocks are some of the most able people around and they also have some of the better improved pasture available, not to mention the correct buildings for inwintering.

Most people go into smallholding as enthusiastic amateurs so the skill required to handle high density flocks is not there; true it can be learnt, but mistakes can be expensive along the way. More important is what the land can tolerate. Even well-drained improved land would be hard pressed to support more than four ewes to the acre during winter and most smallholdings consist of land anywhere from average down to bog and heathland. A more honest stock density for the enthusiastic amateur starting in smallholding would be nearer two ewes to the acre wintered out or three inwintered, but in wintering brings its own problems.

So when calculating profit from sheep you could use figures anywhere from seven down to two ewes to the acre and the lowest is the most honest and hence the one you should use unless you can prove otherwise. To continue with the sheep briefly, you need to know your lambing percentage. It is easy to find figures approaching 200 per cent reared; ie for every ewe you own you would have two lambs to sell, but if 150 per cent is obtained regularly that is good. With poor to average grazing land and a hardy sheep such as the Welsh mountain 125 per cent is nearer the mark and is much easier to manage for the first few years.

From the above figures we will see that a wild variation in performance can be calculated from a particular acreage. If we work on a five acre block and the three levels of performance:

1. Pure fantasy (for the novice small holder)
   7 ewes per acre lambing at 200%
   7 × 2 × 5 = 70 lambs per year
2. Good average and something to aspire to
   4 ewes per acre lambing at 150%
   4 × 1.5 × 5 = 30 lambs per year
3. The level likely to be obtained at first
   2 ewes per acre lambing at 125%
   2 × 1.25 × 5 = 12.5 lambs per year

Assuming £30 per lamb and ignoring overheads (which would be greatest on 1 and least on 3, thus decreasing apparent extra profit) this would give a gross income as follows:

1. £2100 per 5 acres - £420 per acre per annum
2. £900 per 5 acres - £180 per acre per annum
3. £375 per 5 acres - £75 per acre per annum

It is easy to convince yourself of a good income if you use all the maximised figures as used in example 1, whereas the reality is likely to be nearer example 3. The truth is that it does not take much bad luck (inexperience) to turn a small profit into a loss; it has happened before and it will happen again. If you have not banked on the highest profit, it will be possible to ride the loss and learn by your mistakes. Otherwise you may find the bank less helpful next time. Keeping the bank manager's confidence is very important; once lost it is very difficult to regain. It is not unknown for banks to foreclose on people who have a barn full of expensive stock because of lack of confidence. If you say you need an overdraft for a month to cover expenses until you sell some stock, in one month that stock must be sold for *more than you owe*.

The financial plan

In its simplest form your initial plan can be reduced to something similar to: Capital = Cost of property + Repairs + Set-up and running cost + Housekeeping + Emergency fund.

Capital

It is difficult to discuss capital in general terms because everyone's financial situation is different. Some will sell a property and have enough to buy their smallholding for cash, set up their various enterprises and still have enough to invest. Strangely, this group are at greatest risk, having the most to lose as it is easy to dip into that extra capital for day-to-day living. The other effect of having surplus money (unless it is sufficient to provide a private income large enough to support the smallholding indefinitely) is to mask any losses incurred and to remove the necessity of making a profit from what you are doing. As this money dwindles it is very difficult to replace because it has probably not been earned by daily work, but by the ownership of an appreciating property. At times, property prices in parts of the country have risen in one month more than a man could earn by agricultural work in one year, so if you have spare capital take financial advice and invest it *safely* where it can be reached only in an emergency, ie to stop you losing your home.

A safer situation in some ways is to spend most of your money on the smallholding, hence keeping it linked with the property market. This can be arranged by buying the best that you can afford, still leaving yourself enough to finance the setting up and running of the home and business until you are into profit. Purchasing the best you can afford does not necessarily mean an immaculate property needing no work, rather the one with the best potential. If it needs some repair work, remember that will not only cost money, but time as well, time that you could have used starting your business. If you are in the position of buying the smallholding outright and keeping some money for initial expenses and starting the business, this is the best way (other than having a private income, when you can play at it). You still have to meet the regular bills for gas, electricity, water, rates etc, but you can economise in most, except rates. The danger here, if you want some extra cash to finance a project, is the temptation to secure it against the property. Suddenly you are in a different ball game. Now you have monthly repayments to be met, you are at the mercy of interest rates (which inevitably go up); in short you have lost your independence. If you owe nothing to the bank you are your own master. It is advisable to clear all credit or other outstanding loans before you embark on smallholding, although difficult to do when you want all your spare cash for the project in hand; it will be almost impossible later.

The third way of purchasing your smallholding is by using partially borrowed money. This is probably the least desirable route into smallholding but the one many people will have to take (including the authors). If you wish to buy a house with a paddock it is likely that you will be able to raise a conventional mortgage on it, assuming you have a job which provides a suitable income. However, if you wish to buy a larger holding, ie one that is able to add something to your income, most high street building societies will inform you that they do not deal with agricultural properties. After that the banks are your next best option, but you have to show them that their money is going to be safe. What the manager will want to see is a forward projection outlining your plans for the property, initial costs, running costs and profits (you hope). This will be dealt with in detail on pages 28–33. Remember, the loan is secured against your home and land; if you are borrowing less than half the market value of the property the bank is on a safe bet, and however much bad press a foreclosure and forced sale will get the bank, it will certainly get them all *their* money back if not all of yours.

First-time buyer. If you are a first-time buyer, it will be very hard for you, but the potential reward is just as good.

The only ways open to you are either to buy a derelict property and rebuild it or buy a field and try to get planning permission for a dwelling.

Your chances of finding a derelict property within the price range of most first-time buyers is slim, especially as developers and builders are looking for them; however, it is not impossible. The more likely way is to buy some agricultural land without planning permission, and then ride the storm of obtaining planning consent. People have been known to put a caravan on site and hope that no one bothers to complain to the council. (We don't suggest this is the ideal method.) The one hope in all this doom is the recent relaxation of planning laws in rural areas which may make it easier to obtain permission for a small residence.

Housekeeping

You probably now have an idea of your available capital and, barring violent fluctuations in the house market, this is likely to stay fairly stable while you are house hunting. Knowing your capital you must now set about calculating how much you spend to live. This will include everything from toilet rolls to car tax and cornflakes to petrol. Do not fall into the trap of assuming milk, cheese, vegetables and all meat will instantly be produced free and in suitable quantities. These things take time. The longest you should plan to support housekeeping from original capital should be six months, though three months is better.

Set-up and running costs

Next you must calculate the set-up and running costs of your business ventures and how long it will be before they show a profit. On the agricultural side you will not see a payback in less than one year (if you are lucky) and so will not be into profit for two years - hence the need for non-agricultural ventures.

If you are planning a venture involving motor mechanics the chances are you will already have a set of tools. This is true for many hobbies/trades, but remember that when you start using those tools on a daily basis rather than just at weekends they will wear out much faster, and when you come to replace them it may be advisable to do so with professional tools rather than ones made for the DIY market. This, of course, adds to the bills but it is better to calculate the costs now rather than have something break and find you cannot afford to replace it. The time you plan to support your business should be as short as possible, between three and six months being likely - six being the maximum and three the ideal. If you are still supporting the business after six months, you will have to be looking at large profits to make it worth while. Be very self-critical and make sure you are not financing a lame horse.

The emergency fund

The emergency fund need not be very large, in fact it would be a waste of resources if it were, but to know it is there is very reassuring. I suggest one month's housekeeping and business expenses would be enough. You should never need it if you have done your calculations correctly; however, if you have fouled up and something takes a bit longer to come right or doesn't sell first time, you might need it. If you do use it to 'tide you over', it should be replaced as soon as the expected money arrives, otherwise next time it will be the bank manager and he will want interest as well.

Estimating costs

To avoid guesswork when estimating costs for your calculations you will have to obtain genuine prices, both of items you will have to buy and of those you will have to sell. The one which will have the greatest short-term effect is the price obtained for your present property (assuming you have one to sell), but you probably have a good idea of its market value. To estimate the profitability of your business ventures you will have to find out costs of raw materials, take into account overheads such as power consumed by heating, lighting or machinery. Also, less obvious expenses such as advertising and transportation can make large dents in what originally looked like a very good profit. When pricing agricultural goods you will find market reports are useful. These are found in the rural weekly papers and usually give a description of conditions on the day, typically 'a moderate entry of all categories sold to a generally brisk trade. Finished lambs were in particularly good demand.' From this you can judge whether the prices quoted are likely to be representative. The report will then continue with prices:

> Lambs: Light, max. 160.0p; average 147.9p per kilo; Standard, 192.9p, Av. 181.2p; medium 190.5p, Av. 186.9p; Heavy, 165.9p, Av. 162.3p.

Looking at these prices it will become clear that there is a wide margin between maximum price and average price. It is also worth remarking that there are as many prices below the average as there are above, so if you use the average price as your maximum price you should be safe. Avoid using figures from special sales such as Christmas fatstock shows and sales as these attract the best animals and hence better than average prices. However, if you have good quality livestock to sell these are good places as they attract a large number of buyers, especially dealers interested in quality. Other useful sources of data are periodicals such as *Farmers Weekly*. Although aimed primarily at the larger professional farmer, sale prices are still valid and useful.

Obtaining finance

If you decide that you will need extra finance to purchase your smallholding, you may get a rather cool reception from the usually helpful high street building societies. The exceptions are if you are already self-employed, have books showing good profits and intend to continue this trade alongside the smallholding, or if you intend to buy the holding and keep your present job. Unfortunately, these situations can and are difficult to arrange and the chances are that you will find yourselves asking the bank manager for a business loan. If you are asking for this type of loan, the bank can quite correctly ask to see detailed predictions based on figures from reputable sources. What you have to prove is that you can earn a living wage *and* meet their repayments from whatever ventures you choose. You must be ready to defend your figures and answer with confidence any criticisms levelled at them. If you do not have confidence in your plans, how can you expect anyone else to? Rather than try to explain the way we would lay out this document we have included the one shown to our bank when trying to obtain finances for our smallholding.

This is a forward projection showing the inherent profitability of the proposed property.

It is a 16-acre smallholding in West Wales, situated approximately 6 and 21 miles respectively from the Lampeter and Aberystwyth campuses of the University of Wales.

The agricultural projection is based on the production of lamb as this is well documented and easily implemented, though higher returns could be gained from other more specialist products catering for the delicatessen trade of the South and South East.

Beyond the use of the land there is a calf shed, a pig unit and a cattle shed which will be used for extra income.

The costing for lamb production is split into high and low stocking systems at seven and four ewes to the acre. The lower stocking system allows for a lower standard of stockmanship and also less reliability on bought-in food.

The word processing projection is based on experience of part-time work carried out in Poole over the last 18 months, initially for students at the Dorset Institute of Higher Education, Wallisdown and recently for a professional author from London.

The need for educational typing facilities starts after the first term of the second year of any degree course and includes all submitted project work up to and including postgraduate theses.

Work for professional authors is obtained by national advertisement or recommendation and is mainly conducted by post.

Although Dorset IHE, Wallisdown offers only a few degree courses and is situated in an area of high population density, the combination of accurate work and word processor capability brought in a considerable amount of work.

Two university campuses in an area of low population density will provide at least enough for one full-time employment.

Further prospects include purchase of extra word processor(s) and employment of typist(s) working from home. Also extension of services to include audio typing (by telephone on to an answering machine) and photocopying facilities.

There is an eight-berth caravan installed on site. The previous owners apparently had no difficulty in letting this solidly for the full season (July to September inclusive). As this is a very picturesque spot in a popular area we have no doubt of equal success and have therefore used some typical weekly rental figures to calculate income over a 16-week period.

INCOME FROM 16 ACRES BY THE PRODUCTION OF LAMB

(figures obtained from *Farmers Weekly*, 20 March 1987)

Stocking level 7 ewes per acre

Year 1-112 ewes purchased @ £36 each = £4032

Lambing @ 150%* = 168 lambs

Gross margin = £51.61 per ewe
Feed and keep costs = £ 9.21 per ewe
Net profit = £41.69 per ewe

Total = £4670.00

Profit in first year £4670 - £4032 = £638

Following years allowing 15% new blood replacement

£4670 - £605 = £4065

Stocking level 4 ewes per acre

Year 1-64 ewes purchased @ £36 each = £2304

Lambing @ 150% = 96 lambs

Gross margin = £51.61 per ewe
Feed and keep costs = £ 5.00 per ewe
Net profit = £46.61 per ewe

Total = £2983.00

Profit in first year £2983 - £2304 = £679

Following years allowing 15% new blood replacement

£2983 - £346 = £2637

INCOME FROM WORD PROCESSING

Figures obtained by previous experience with local educational establishment and also professional author.

No initial outlay as word processor already bought.

Typing and printing rate 1500 words per hour
@ £2.50 per 1000 words

Allowing 20p per 1000 words for paper and ribbons.

Net income £2.20 × 1.5 per hour = £3.30 per hour

Assuming 40-hour week £132 per week

*This is an upland percentage. A lowland holding should give nearer 200 per cent lambing as is common in Dorset

42-week year £5544 per annum
- 5% maintenance = £5267 per annum

Assuming 25 hours per week £82.50 per week

42-week year £3465 per annum
- 5% maintenance = £3292 per annum

INCOME FROM LETTING EIGHT-BERTH CARAVAN

Caravan is already in position. It is plumbed with drainage, mains electricity and water.

June	@ £75 per week × 4 =	£300.00
July	@ £95 per week × 4 =	£380.00
August	@ £95 per week × 4 =	£380.00
September	@ £75 per week × 4 =	£300.00
		£1360.00

Allow 10% maintenance £1224

No problems are envisaged in letting the caravan as previous owners always had full bookings.

FULL POTENTIAL		PART POTENTIAL
Land	£4065	£2637
Typing	£5267	£3292
Caravan	£1224	£1224
	£10,556 per annum	£7153 per annum

Although parts of this now make us cringe it is easy to be critical with the knowledge of hindsight and extra experience, but basically the main points used in the projection were sound.

The stocking levels of seven ewes per acre were seen through rose-tinted spectacles, but we were surprised how badly four ewes per acre treated the land in practice. One good point is that when the first lamb crop were sold the sheep had paid for themselves and made in the region of £500 profit, which was close enough to the projection to justify the figures.

The word processing has never realised its apparent potential. The figures were all sound but the expectations of work 25 hours per week never materialised. Time and money have been spent advertising, but still the work is very slow in coming. If this had been the only or main income, it could have been disastrous; instead it was only unpleasant.

The caravan prices are about £10 per week too high, but more important than that is the effect of season length on total income. In this case the season length was overestimated by about 100% per cent, giving a true income less than half that quoted.

Finally, the figure under the heading full potential belongs in the fantasy category. That under part potential should actually be headed maximum attainable by these means. At the time we did not understand the true necessity of non-agricultural ventures. However, ignoring the fact that some of it was little more than wishful thinking, it was good enough to borrow 20 per cent of the required capital on a business loan. As we pointed out earlier, at that level the bank was on to a safe investment and even if the plans failed totally, they would get their money back.

Making the money work hard

Whatever you decide to do you must ensure that the money invested is earning at a realistic rate. It is senseless to invest large sums of money in a project, then work hard at it both physically and mentally, only to obtain an income similar to that obtainable by investing the money in a building society. It may sound ludicrous that anybody would do such a thing, but with some agricultural ventures run on a small scale you would have done well to see any profit at all. It is all-important to get the money moving as soon as possible. Therefore your prime objective should be to start the short-term payback schemes first and let the longer-term projects come on later. In this way you will protect your capital. It is absolutely essential that you monitor the flow of money as it passes out and then returns. You at least must know exactly how much any one project is making or losing at any time. This is the only way that you can direct the flow of cash in the most profitable direction. It is sensible at first to put all money through the business account, as this keeps things neat and tidy and allows you to see the cash flow simply by looking at your cheque book and your paying in book.

Having identified the projects between which your money is to be divided, you must set targets for these projects. Initially these will probably be time scales for setting up and general organisation, but financial targets must also be set. You should aim for a given payback period and then strive to meet that target. You should also set yourself work targets. It is easy to allow yourself to do very enjoyable or at least untaxing work at the expense of jobs which earn you money. Until you are established you must put the financially rewarding jobs first. The more financially rewarding a job, the higher priority it should have. Beware of over-stretching your finances. You may have an idea that promises to make you a good profit. If you put all your money into it, you place yourself in a very vulnerable position, especially if it concerns livestock. The dangers of gambling all on livestock are many, not least that animals can die sometimes for no apparent reason. If you have spent all your remaining capital on some calves and they scour and die, you are left with nothing, whereas if you had spent only part on calves the loss would not have been total, giving you another chance. The other problem with livestock is that they eat all the time you have them, and if the market prices drop to a level that offers you an unacceptable return on your stock, you have to make a decision; either to keep them and carry on feeding them (hence spending more money, daily increasing what they already owe you), hoping the price will increase again, or to sell them and cut your losses. Inanimate objects do not involve these problems unless they are bought on borrowed money, in which case the interest has an effect similar to the feed bill; as a car or a walking stick will have the same value from one week to the next and will not eat anything (or die) while you are waiting for it to sell, this gives your investment some kind of security.

Professional advice

What you should do is simple - take it. If you are unsure about running a small business, there are courses available. The same is true for bookkeeping which can save a lot of money later on. If you are unsure about the best way to organise your business, ie sole trader, partnership or limited company, again take advice. This is something you could discuss with your bank manager (after he has agreed the loan!) or possibly with an accountant as there may be tax benefits to be obtained by certain arrangements associated with your particular circumstances. Your decision regarding the type of trading will then further affect the type of professional advice you will need to take.

Sole trader

As a sole trader you do not need to register. You alone are responsible for the business bills and any other liabilities. However, as there is little differentiation between your personal finance and the business, your savings and property are at risk should your business fail and accrue debts. This should not be a great deterrent as with careful handling and constant reviews of the financial situation, serious debts should be avoided. Sole traders are taxed on their business income (ie profits) so it is essential to keep accurate records for presentation to the Inland Revenue.

The majority of people starting small businesses do so as sole traders: it is simple and, barring major cash flow problems or financial crises, is the best way for the smallholder.

Sole traders and partners pay Class 2 and Class 4 National Insurance contributions; the figures for 1989/90 are: a flat rate of £4.25 per week for Class 2; 6.3 per cent of profits between £5,050 and £16,900 a year, (a maximum of £746.55) for Class 4.

Partnerships

A partnership within a family group, ie husband and wife, is the most common one found in smallholding. This can work well as with complementary talents there is greater scope for maximising profits. As the money generated from this type of partnership goes into one household there is often no formal legal partnership agreement drawn up by a solicitor. If, however, the partnership is outside the family group and the profit (or loss) has to be split between two or more households, a legal document should be drawn up stating the type of partnership, equal or otherwise, and laying down some basic ground rules for the running of the business.

Partners in business should be chosen with great care because each partner is responsible for the liabilities of the partnership. It is therefore worth consulting a solicitor about the necessity for an agreement between the parties. It would spell out, among other things, what happens if a partner wishes to leave or dies, what each should contribute and what each should be able to draw, and the division of profits. Partners are taxed as sole traders.

Limited company

In a limited company the liability of the individuals concerned is limited to the face value of their shares. If a person holds 90 out of 100 £1 shares in a limited company, he cannot be compelled to pay off more than £90 of the company's debts whatever it owes, unless he has given personal guarantees. The company is a clearly defined legal entity owned by its shareholders and controlled by its directors (usually the same people for small businesses), and trading according to certain articles of association which have been drawn up first.

A problem is that the home and smallholding are so intimately associated that you could not easily form a 'Smallholding Ltd' to protect your home. A business run *from* the smallholding could be a different matter, and a solicitor could give you the information required to protect your home. An expense with a limited company is the way the books have to be kept and audited. Annual returns have to be made to Companies House, thus making your balance sheet public and increasing your accountant's bill. An 'off the peg' limited liability company can cost less than £100, but for most smallholding ventures, being relatively small and likely to change to follow the market, it is unnecessary and possibly undesirable to go limited.

Before you do anything definite it is worth consulting a solicitor and an accountant. Explain your plans, your financial situation and your projected turnover. Sound advice at this stage can save money and aggravation later on.

Chapter 3

Choosing Your Smallholding

Choice of area

Your choice of area will depend on a number of factors. One of the most important will be property prices. These vary widely from area to area and large variations can be found within an area. Prices depend upon immediate surroundings, condition of land and buildings and how keen the vendor is to sell the property. So if you are choosing an area primarily on property price, be sure to look at a good sample of available properties.

Another main consideration is the local climate and its effect on your chosen agricultural ventures. If you want to raise early soft fruit, don't go to the north of Scotland. Obvious I know, but there are other more subtle climatic variations. Rainfall is greater in the west of Britain, for example. If climate is likely to affect your choice of area, visit your local library for information about rainfall, sunshine and average temperatures. However, if you intend to keep the smallholding to help with the housekeeping and earn your money in other ways, you can easily adapt what you do and the way you do it to fit the climate. It is difficult to comment on the various areas and their pros and cons because individual tastes will favour different things and prices vary greatly. In general the south of England is the most expensive and has the most pleasant climate. As you go north property becomes less expensive and the climate becomes more severe, although there are micro climates within areas often along coastal plains.

It is likely that you will have some idea of where you would like to settle, but if you do not have a look through the property adverts in *Exchange & Mart* and *Farmers Weekly*. These should give you some leads and, if you know little about a particular area, again the library will probably provide more information.

A good start is to go to a main post office or library where a full set of Yellow Pages is kept and, choosing those that cover your selected areas, copy out a selection of likely estate agents. You can either telephone or write to them. The phone call is useful as it gives you immediate feedback. You will know the size of property you would like and your maximum price; they will tell you if you are likely to find anything like it in the area. If it sounds possible, request details of suitable properties.

The details sent from requests like this will give a great deal of information about the areas from which they come. You will find various business properties for sale such as kennels, organic nurseries, rabbit units, free range poultry etc. Ask yourselves why are these for sale as 'going concerns'; there may be a very good reason or it may be difficult to make them pay because of location or transportation of produce.

If you are uncertain about the best area to begin with, you will build up a picture of what each area offers as you receive details from different estate agents. This, coupled with any background reading that you have been doing, will gradually bring you to favour one or two areas above the others. This is the time to visit your prime area. Possibly it is best to do this with minimal planning concerning accommodation. If you stay in a bed and breakfast this throws you into contact with the local people and it is from them that you will learn most about the district. It is a good plan to visit the estate agents with whom you have been in touch so that they put a face to a name. It also shows them that you are looking seriously for a property in the district. It is well worth enquiring whether they have anything new on the market because the best buys are sold almost as soon as they are offered. It is worth visiting a couple of likely properties just to compare them with the estate agents' descriptions; this will give you a frame of reference when reading future literature. Estate agents vary greatly when describing property. If most of the land is steep, some will say 'generally sloping land', while others will say 'level and some sloping land'. Both are the same, but one produces a different picture in your mind. If a description says 'steeply sloping land', you will probably have to go up it on your hands and knees. Try to drive around as much as possible. Drive down minor roads just to see where they go. This combined with the local ordnance survey map will soon give you an idea of the area. Stopping in villages and towns to walk and get a feel for the place will help you to decide if you could be happy there. The purpose of the first visit should be to decide whether the district 'feels right' for you.

Choice of property

Once you are happy with an area concentrate on finding a suitable property. You will know what you would like and what you can afford; somewhere between these two figures lies your new home. The asking price is usually higher then the expected price and on some smallholdings can be in the region of 20 per cent, even 25 per cent, more than the vendor will accept, therefore allow yourself to look at property up to about one and a quarter times what you can afford. If you find one of these is suitable you can always put in a low offer and hope for the best.

Many estate agents now send printed lists of properties instead of individual information sheets in the first instance and, as these may cover a period of months, it is worth telephoning to check if a property of interest is still available and how long it has been on the market. (If the estate agent will not tell you someone in the local pub usually will.)

When deciding what constitutes a suitable property you must consider potential. If you want a certain size house with a certain amount of land, the chances of finding the ideal property are slim. If you buy a property with the right size accommodation but little land you may have to wait a long time before a neighbouring field comes up for sale; conversely if you have enough land but the house is too small, the only considerations are planning consent and the money to do the extension. When considering out-buildings the converse applies. They are expensive to put up compared with their value when part of a property; if you need out-buildings for your ventures try to buy them with the property. Another consideration must be the location, whether you want a village property or one that is isolated. The idea of solitude is very appealing on a warm sunny day, or when you are surrounded by noisy neighbours. The problem comes with bad weather, unreliable transport, even running out of tea unexpectedly. It is true that few places are really isolated when you have a car, but if you are a mile down a gravel track and then five miles from the town or village, taking the children to school every day will become time-consuming and expensive. Nearly every property about which you receive details will have something going for it; equally they will nearly always have something which detracts from their appeal. You have to sift through these; it is doubtful whether you will be able to go and see all the properties which might be suitable. A points system is one way of sorting them, possibly giving marks out of ten for such things as amount of land, out-buildings, number of bedrooms, size of kitchen, location and so on. The other way is to select those that 'feel right'. Of course, it is another thing when you see them with your own eyes; what sounded ideal on paper and even looked good in the photograph can turn out to be a decaying shell surrounded by boggy land and fronting on to a very busy road. You have to go and look and, if it is no good, put it down to experience and move on to the next one.

Something we feel is important when viewing a smallholding is potential; this is also very difficult to describe. The level of potential depends greatly on what you wish to do with the property. What will be regarded as a useless boggy area full of rushes and springs to one person will appear to be the ideal location to dig ponds for fish farming to someone with different aspirations. Generally, a property with potential is one that can be improved or changed so as to increase its value/desirability above that of its original condition.

Viewing property

When you have your batch of chosen properties in a given area you need to go on a serious viewing trip. It is best to phone the estate agents in advance to ensure the properties are still on the market and to arrange appointments for viewing. Try to plan your day(s) to give maximum viewing with minimum driving, ie one district one day and another the next. Allow yourself plenty of time to look around and also to discuss the property afterwards. It is worth making a few notes to help you remember what you have seen where. To avoid being overwhelmed by various properties it is advisable to look at no more than four per day; this allows time to chat to the vendors and thus find out far more than the estate agents can tell you about the immediate surroundings, amenities, neighbours etc.

Winter is arguably the best time to look at smallholdings because any problems associated with water (leaking roof, rising damp or badly drained pasture) become more apparent at that time. Also properties don't sell as fast then as in the summer. However, against that there are possibly not as many properties on the market in winter. When looking at the dwelling house pay special attention to the roof and floors, both of which can be expensive to replace. Take note of any damp patches on the walls and ceilings. If there are such things as a solid fuel range or kitchen units enquire if these are included in the price; even if they are not to your liking, every little helps when you first move in and you can always sell them later on.

The out-buildings need careful scrutiny. Look for drainage. Do they have concrete floors? Do they have any hard standing outside them? Which way does the water run when it rains? All these things can affect the usefulness of a barn. If, for example, there is no hard standing in front of a barn any movement of livestock in and out in wet weather soon produces mud problems; the same is true for vehicles if you use the building for storage and want to load/unload at frequent intervals. Some hardcore is the minimum required - more expense if it is not already there. The general requirements for a barn are that it should be dry and airy without being cold and draughty, although anything causing dampness will be a problem. You are unlikely to find the perfect barn, so be prepared for some maintenance.

The land is far more difficult to assess than the buildings. Its appearance will depend upon its management over the last weeks, months or years. If it is full of rushes or there is a lot of moss in the grass, this indicates wet acid land. Look at the ditches: if they are badly blocked or non-existent you could have found at least some of the problem. If the ditches are there and are full of stagnant water, look to see what has caused the blockage. If the blockage is on someone else's land, there could be a long-term problem with drainage. Also check the water supplies.

Another important point is the depth of soil. When the soil is very shallow with underlying rock the ground may be prone to water logging in winter and drying out in summer. The only way to satisfy yourself about soil depth is to take a garden fork or steel bar with you and probe the soil at numerous places around the holding. It would be polite to ask permission to do this, but if the owner or agent refuses you can only assume the worst: very little soil.

Look at the fencing; you would hope to see three-foot sheep fencing with one strand of barbed wire above supported on pickled stakes. The wire should be tight and the stakes firm in the ground. If the perimeter fence is like this you will have few problems, otherwise allow 70 pence per metre for new fencing excluding labour.

Consider general location. How exposed is it to the prevailing wind? Is there any shelter for stock? Is there a river likely to flood the land (or property)? Is there anything close by which may cause you problems in the future? This could be anything from an adjoining farm doing early milking to an abattoir and the various aromas they produce. Remember, such things are part of the district. If you move in next to them it's no good complaining later. No one likes it when a newcomer to the district starts complaining about something that has been there for years, especially if it also provides local employment. Make sure you are satisfied with the property, the surrounding environment and all that goes with it before you make an offer.

Property in need of renovation

If you are looking for a property which is in need of considerable work you may find that it is eligible for certain grants towards this.

The grant system at present is divided into three types: the intermediate, the repair and the improvement. All three are based on the rateable value of the property.

The intermediate grant
This is for basic amenities such as bath, toilet and hot and cold running water in the house. It generally covers cases where the amenity was previously lacking, but can be for repair work under certain circumstances. Some roof repairs can come into this category, but consult with the local council. If the property does not have the basic amenities, the grant is automatic but may be subject to a waiting list. As with everything else there is only so much money to go round.

The repair grant
This is for properties built in 1919 and before. This covers more structural items such as roof, gutters and windows, and so is probably the most useful of the three. This is reflected in the fact that in many areas there are waiting lists running into years (three years in our area).

The improvement grant
This is to bring a basically sound property up to a higher standard; for example, fitting larger windows. Again there may be long waiting lists, so enquire from the local council involved before banking on getting any assistance.

There is a leaflet available from your local council housing department describing all these grants and how they apply to different cases.

It is also worth noting that, as with many things, changes are in the pipeline and these grants are no exception. The most important change is that their availability will be on a means tested basis.

It should be borne in mind that the property needs to be basically sound before you start improving it, and unless you know what to look for take professional advice. A simple crack may be just some old plaster or it could indicate that a wall is about to fall down. Serious structural work can prove very expensive and if you get involved with new roofs or underpinning walls, what looked like a bargain at first can soon cost you well over market value.

Land in need of renovation

The potential of the land associated with your chosen smallholding will vary greatly with area, but however good it is there will always be room for improvement. If the pasture has been badly neglected a complete programme of drainage, ploughing and re-seeding may be in order. Often, however, this does not need to be done to improve the land sufficiently to make a useful improvement in its stock-carrying potential. It is natural and sensible to aim for the best return from your land, and this can be done by taking on a larger area of land or by improving a smaller acreage. The stock-carrying potential difference between old poorly drained pasture and well-drained newlay is in the region of 300 per cent. Another way of looking at this is that, for every acre of old pasture you have, a similar improvement in output could be achieved either by buying a further two acres of similar land or by draining and re-seeding what you already have.

Land prices vary greatly with area and time, but in general land will always be a good asset on our overcrowded island and at present average pasture is valued around £1000 per acre. (This could be as little as £500 per acre for very poor land in some areas or more than £3000 per acre for prime land in more fortunate areas.) Therefore it is possible to make the generalisation that to treble your output by purchasing land of a similar quality would involve some £2000 per acre already owned.

The expense involved in improving the land which you already own depends on its present condition. If it is covered in scrub and brush wood, there is extensive work involved in clearing and removing stumps before a plough could come anywhere near it. But let us assume the land is not that bad. Wet land with rushes, mosses and weed grases can be ploughed directly, but unless the drains are in good order ploughing and re-seeding will not give full benefit. If the land is already drained but still holds water, it is possible that the situation can be improved by mole ploughing. This draws a special plough through the soil leaving holes (similar to mole runs) behind it. To do this a powerful tractor and, of course, the mole plough itself are needed, but an indication of cost could be obtained from an agricultural contractor. The problem may lie in the state of the land drains themselves. If they were laid years ago the chances are that they are not laid at a very great depth and the trench was not back filled with stones. Also they were designed to take the weight of horses and horse-drawn equipment; as time has gone by agricultural equipment has become progressively heavier, compacting the soil and possibly cracking the old clay pipes. A broken or blocked drain can show itself as a darker green patch in a field or a boggy patch or strip aligned with the run of the drain. Let us assume the land is poor pasture, but needs redraining and re-seeding.

To lay drains it will be necessary to employ the services of a contractor as he will have all the machinery required, but it may be possible to save money by labouring yourself.

This is a list of costs per metre incurred laying drains (1989):

open up	50p
cost of pipe	35p
cost of stone	55p (approx)
stone fill	20p
back fill	15p
Total	175p

Here we have a total cost of £1.75 per metre for laying drains. An acre field 70 yards square would need drains every 20 yards if it was either very wet to begin with or needed to be well drained for some other reason. This would give three runs 20 yards apart giving a total of 210 yards of drains for one acre costing £368; if a lower standard of drainage was required only two runs could be laid at a cost of £245.

The ploughing and re-seeding should also be done by a contractor who has experience to back his judgement of what is required and when it is best to do it.

This is a list of costs per acre for re-seeding.

cultivation	£50
seed	£25
fertiliser	£35
lime	£30
Total	£140

This gives an idea of the cost per acre for re-seeding.

Combining the drainage cost and the re-seeding cost gives

about £500 per acre which compares favourably with the purchase of extra land. Consider a five-acre holding of average pasture: to treble output by land purchase would cost in the region of £10,000. However, by land improvement this would be closer to £2500. Remember that the land has to be stocked and the extra stock managed whichever way is used. Points to note are that there is no guarantee of the re-seed taking first time; as with all agricultural pursuits you are at the mercy of the weather. Take *local* advice on the best time to re-seed. Also a re-seed does not last forever and you will still have to maintain the land by having it mole ploughed every three years or so and keep ditches free.

The asking price

A smallholding is basically a house, some out-buildings and land. From this description one could expect to be able to calculate the price by the simple addition of respective values. This, however, never reaches the actual value because a premium is required for the three items coming together. Land also tends to be more expensive in small plots than in large areas, the reason being supply and demand. There are more people with the ability and inclination to buy one or two acres than there are those wishing to buy 100 or 200 acres. Out-buildings add to the desirability of a holding for without them many ventures, both agricultural and otherwise, become difficult or impossible to undertake. The individual value of the dwelling house and the land can be easily estimated by comparison with other properties and accommodation land advertised in the area, whereas the value attached to the buildings depends not only on their quality and condition, but also on how versatile they are. The final decision has to be made by asking 'Is this what we want and does it fulfil our needs?' If the answer is yes then, finances allowing, you may be right to pay a premium to get the property that suits your needs.

A problem can occur when an unsuitable property is found at what appears to be a bargain price (though this is a rare occurrence with today's inflated prices). However, there has to be a reason for the depressed price and it will probably be that if it is unsuitable for you it will also be unsuitable for many other people. You may convince yourself that it only needs modernisation and an extension to make the house acceptable and the old rusty iron barn would be serviceable if re-roofed and the floor concreted. The land, of course, could be brought into a very good state by drainage and re-seeding. Such a bargain would almost certainly prove a disaster. The amount of time and money required to put things right, so that you could start to earn a living, could be enough to drain not only your resources but your enthusiasm as well. It is better to pay more cash 'up front' and start earning sooner than end up paying at least as much and losing time on your paying projects. Of course, there are real bargains, but most of these are bought by builders or professional developers who have cash available to make fast closures on the deal. If you are in a position to compete with these people, go for it. Cash talks, especially when the bank manager is threatening to foreclose.

When considering the property valuation against its condition, there needs to be some allowance made for all shortcomings so that, if you moved in and rectified them, the value of it would have been increased to cover your time and money and also give a profit on top. Many of the asking prices do not allow this margin.

When you have decided on a price that you would be happy to pay for a property, make your first offer lower than that; you can always increase, but it is very difficult to decrease your offer.

Be aware of exactly what is included in the asking price. If there is some machinery or stock on the holding which is not included this could be used to overcome a deadlock. If you have made your 'final offer' and they have made their 'final concession', rather than lose the property it may be possible to agree to their price if certain items of stock are also included in the sale. It may not have the value but it saves face on both sides and allows the deal to proceed.

Before you make the offer think of the suitability again. It may seem perfect at the time, but consider the changing seasons. Also consider the passing of years; if you are young, would it be suitable for (more) children? If you are not so young, would it be convenient for shopping without transport?

Think again of your intended methods of earning a living; if they should fail what would you fall back on? Would it still be suitable? Location can play a large part here; road frontage may seem undesirable, but if you wish to do farm gate sales, pick your own, or bed and breakfast, a mile-long track could deter your customers. Buying a smallholding is different from buying a house; you want not only a place to live, but also a place to work. They must both be right for the whole thing to succeed. When you feel it is right make the offer and get ready.

When you make your initial offer the estate agent will probably want to know your financial position, and assuming you have a property to sell, how far advanced your sale is. You want to be able to say 'my property is under offer and we expect to complete in x weeks' and if you can say this truthfully so much the better.

The search

Once an offer has been accepted and the purchase placed in the hands of your solicitor, the search will be conducted. This should bring to light any problems associated with the land or property such as impending development or a new bypass; this is standard with any property. With a smallholding there are other considerations which you may wish to discuss with your solicitor. There may be rights of way across the land or possibly you may only have a right of access across someone else's land. It is worth checking on any restrictions imposed on you or by you on others. If you are looking for seclusion a right of way through your back garden could be rather annoying. Maintenance of fences and ditches is another point to raise; this should be stated on the property deeds and is worth knowing for future reference. However, if a fence is down or a ditch is blocked and it is giving you problems, the sensible thing is to rectify faults yourself and not worry about who is responsible.

Wayleaves are often encountered in rural properties where it has been necessary to run a service, ie mains water, across another person's land; for this privilege a fee has been paid and an agreement signed giving access for maintenance, but also guaranteeing to compensate any damage caused. There may also be conditions associated with the wayleave; for example, the service may only be used for agricultural purposes. This is unlikely to pose any kind of problem, but you should be aware of the possibility. This fee is normally a lump sum paid on granting of the wayleave and therefore, excluding problems, should cost you nothing if already arranged. If power lines pass over your land and you have poles or pylons situated within your boundaries, the electricity board pay you a wayleave fee annually. This varies with the type of pole and the use to which the field is put. Thus you are paid less for a single pole than a pylon and less if it is in rough grazing land than arable. For a single pole in a grazing field the payment is about £5 per annum.

In general take the time to discuss and understand any points raised by the search or anything in the deeds that you do not understand or that your solicitor points out. Take your solicitor's advice, as what may appear to be a problem to you may be perfectly normal practice or vice versa.

Selling your property

This section states the obvious, but sometimes it is well worth stating. When selling anything presentation is all-important. The difference between the tidy garden with a few flowers around the front door and something less attractive can cost thousands of pounds. Most prospective buyers will drive by in a car first. If the outside appearance is not inviting you will probably never get them inside. If the view from the road is not very inviting spend a bit of time and thought on brightening it up a bit; some large pots with flowering plants always help and you can take them with you. Decorative order is a great selling point. If a property has a light, clean feel to it, it will have more appeal than one in need of a lick of paint. It is well worth spending some time and money to improve the 'feel' of the place.

When it is time to place the property on the market it is best to have two or three agents (this from personal experience) as a sole agent does not have the same incentive to sell the property actively and can let it sell itself. Make a few phone calls to obtain good percentages; remember half a per cent can equate to many hundreds of pounds. Ask what you will get for your money. Will your property be advertised in the press? Will they use a photograph? Make them work for their money – you will have to pay enough.

Moving

After much nail biting, many sleepless nights and agitated phone calls the deal will come together, contracts will be exchanged and you will have a date for moving in. This can happen rapidly towards the end so be prepared, know what is to be taken and pack first what you can do without for a short while. If you decide there are things that you will not need in your new life sell them, but be sure that they really will be unnecessary before you do so. When it comes to moving the cheapest way is to hire a lorry and, with the help of your strongest friend, move yourself. Find out hire charges in advance and compare them with removal firm quotations. It really depends upon how far you are moving and how much you have to move, but, assuming the largest van drivable on an ordinary car licence is used and you are not doing more than a 500 mile round trip with shared driving, a 24-hour hire should be sufficient.

In some areas it is necessary to inform the electricity board of your arrival in writing two weeks in advance. This is particularly important if the property has been vacant prior to your purchase because the company fuse will have been removed and there is nothing worse than not being able to have a hot drink upon your arrival. Because we had not given sufficient notice we had to make do with a quarter mile extension lead from a neighbour's barn for three days. It is useful to know where the main water stop cock is and also the meter, if there is one. If this is not obvious outside your property it may be on the roadside where your main joins. If your supply crosses fields to get to you (see wayleaves on page 46) it may be some way off. However, the water board should be able to tell you if given sufficient notice.

Owning a smallholding does not make you a smallholder: running it and making a living from it does that. It is important that at all times you do not over-stretch yourself. The difficulties encountered in smallholding increase in proportion to the money owed. When choosing and buying your holding large sums of money are involved and a small percentage saving at any point can possibly put thousands of pounds in the bank. If you enter smallholding owing as little money as possible, it may make setting up your business more difficult in the short term, but it will help your finances in the long term. If every pound earned can be ploughed back instead of being split between you and the bank, the business will grow much faster after the initial set-up stage.

Chapter 4

Getting Started

Realisation

After the turmoil of the move you wake up one morning and find that what you have planned for, striven for and dreamed of has finally happened. You now have a smallholding. For the first few days there will be plenty to do just finding the teapot and generally trying to create some kind of sanity out of the chaos.

Before you start anything major you need to get to know your new home, both the house and buildings and also the land. There may be essential jobs that you knew about before moving in, such as a leaking roof or broken window. There will undoubtedly be other work required which was not so obvious. Perhaps fences need to be mended - possibly the wire is still good but a post has rotted off in the ground. Such things may not be obvious until you actually walk around the fence and check each stake before an animal leans against it. (Stock have an uncanny knack of finding the weakest post and leaning against it.) Try to be organised from the start. Carry a small notebook and pencil with you if necessary and jot down everything that needs attention: ditches to be cleared, hedges trimmed and so on. With all this information in one place it can help to plan work. If a digger is needed for one job you may find other jobs for it while it is on your land.

Something encountered by newcomers to smallholding and self-employment in general is a lack of self-motivation. Although there will be jobs that require urgent attention, many can wait. It is easy to go to the market 'just to see how the prices are' rather than do something more constructive. But protracted tea breaks, late starting, early finishing and in bad weather not starting at all are a sure way to bankruptcy.

Those who have previously been self-employed will fully understand how difficult it can be to make money - make being the operative word. Others who have always been employees may not fully understand that when you are not working you are not earning. If you lose one hour a week every week it would cost you one week's wages in a year and that is only 12 minutes a day. This can soon turn your small profit into a large loss. Of course, there will be things to do that do not make you money directly (property improvement is a good example) but if it makes life easier or increases efficiency the profits will be indirect.

Smallholding is a way of life that must be enjoyed to make it bearable. It is not like a job that can be borne eight hours a day by thinking of the pay packet at the week's end. For this reason it is unwise to embark on any project which you find unpleasant either as a whole or in part. No one actively enjoys the time when lambs go to slaughter, but if every time you see your lambs you dread the day they are ready to grade this will take all pleasure out of the enterprise, and it would be better if you found something else to do with your land.

If you do not become too familiar with your 'farm animals' the pet syndrome does not develop; this is a good reason for keeping pets and farm animals segregated in your mind.

Division of work

Before long you will have compiled a good list of jobs requiring attention. Somehow this list has to be graded so that you can not only do the most important work first, but do the jobs in a logical order, each one building on another, rather than randomly doing a bit here and a bit there. The other advantage of having a list is that when you finish one job there is no time-consuming thought required before you start the next one, just a glance at the list.

The initial grading can be done under the headings:

- Essential, ie leaking roof or no water.
- Important, ie starting the business to get the money flowing.
- Fun, ie whatever you want to do most.
- The rest, ie all those jobs that will get done one day.

The essential jobs obviously need to be done first. A new fitted kitchen should not be in this category as a couple of old tables would do just as well for a short period. Be prepared to rough it for a while and the luxuries will seem so much better when they come.

Essential jobs are those which will get worse if not attended to, those which cause actual physical hardship or present a danger, and those which cause problems with earning money. If the workshop roof leaks or the floor is mud, get it fixed before you start work, otherwise tools will be ruined and items dropped will be lost for ever resulting in bad work and short tempers.

Once you have a safe and healthy environment you must start

to earn the money. This is the most important category and will be dealt with in the next section.

Having fun must be taken seriously because 'living on the job' as smallholders do it is easy just to work; this can be as destructive as not working enough, only this time family relationships suffer. Take time out with your family to do whatever you fancy. Sunday afternoon is usually a good time, or late evening in summer. These breaks allow you to appreciate what you have worked so hard to attain and should be very special times for you.

All work under the heading 'the rest' will probably have to wait until the business ventures are up and running and into profit. The very fact that they were put in this category indicates that they are not essential, they do not earn money (therefore they must cost money) and they are not fun. So let them wait until either you can afford the time and money without anything else suffering or they are promoted to a higher category.

Getting the money flowing

There will be a time while you are settling in and doing the essential work when making money will not be high on your list of priorities. But this period must be kept to an absolute minimum. All the time you are not making money you are losing it. Your capital, whether cash or material belongings, will only last so long. This 'wealth' which you brought with you is precious and should be conserved. Money, especially in large quantities, is very difficult to earn in rural areas and many people who start smallholding think things are difficult during the first year to 18 months. But it is only after this period that the goods you brought with you start to need replacing and this puts a more realistic drain on your resources. It can be seen from this that it is essential to start earning within the first six months, allowing time to build up towards the one-year crisis. The more available cash you started with, the longer the crisis takes to occur and the more money is potentially lost in the meantime.

When you survey the projects you intend to undertake try to arrange them in terms of time to profitability. Often those showing a profit soonest will give a poorer return than longer-term investments. But it is pointless going broke waiting for the big one when you could have started two or three smaller projects which would have tided you over until then.

The agricultural ventures tend to be in the long-term bracket, often not paying back capital invested for at least a year.

Medium-period projects are those which include some manual input, making something or buying something to repair and resell. The quickest turnover tends to be made just buying and selling; this could be buying in bulk and splitting the commodity into smaller units for resale with a suitable mark-up, or touring the farm sales looking out for items which you know where to sell at a profit. Beware of the time involved with sales; if you have outlets for various types of item from furniture to old farm implements you could do well, but if you were looking at only one category of item your time could be very poorly paid.

Many smallholders start by buying a wide range of different animals and then spend a lot of time and money looking after and feeding them. They are usually disappointed by the return obtained. It is true that animal husbandry is one part of smallholding that most people look forward to and want to start as soon as possible, but at present it is one of the least remunerative ways of investing your time and money. Once you have developed a profitable business, then you can start your livestock projects when anything earned by them will be a bonus and any loss will be at least bearable. However, if you start the livestock first any loss could prove difficult to recover from and you may find that there is too little cash left to start your more profitable ventures.

Tailoring output to maximise profits

This is particularly concerned with agricultural ventures, but is applicable to most things. Any business venture will require money to start it, time and money to conduct it and will return money when the commodity is sold. A variation in the input either at the beginning or later can make all the difference between profit and loss at the end.

If these costs are fixed by, for example, the cost of purchasing a sow and feeding her, there is little scope for savings to be made. She will cost a set amount to buy and if you economise at this stage you will probably get an inferior animal. It is unwise to economise on food as a badly fed animal will not perform well, wasting the time you spend on it. When the sow produces piglets these have to be fattened (after weaning) and will eat a large (and measurable) amount of food before they are ready for slaughter. All these steps are well defined and leave the smallholder little room to economise.

The only way to increase your margin is to modify the final product to add value to it. To continue with our sow and piglets, one possibility would be to sell suckling pigs for banquets and similar functions. This would save the considerable outlay required to fatten the piglets to porker weight (100–120lb live weight). The problem then becomes one of arranging outlets, ie marketing the product (see Chapter 7). An alternative would be to fatten the pigs and then to cure them for sale as bacon. There is a good market for 'traditional' foods and little difficulty should be found in selling such a commodity; here the problems could come from developing the skills required to undertake such a venture and, of course, the premises would have to be up to public health standards. Sadly, it is rare to get something for nothing and smallholding proves no exception to the rule.

Another example of maximising output concerns egg production. Many people produce free range eggs and sell them to friends or to a passing trade, often holidaymakers. With the number of people doing this the present price is between 85p and £1 per dozen. This compares unfavourably with that asked in most supermarkets and in truth barely covers the cost of the laying bird and her food.

However, with a little thought and planning it is possible to turn this meagre income into a much more healthy profit for a few months of the year. The cost of day-old chicks is far more than the cost of the eggs. Whereas you would receive 7p to 8p for each egg you could expect 70p to 80p for a day-old chick. They would need to be produced early in the year necessitating some artificial light in the chicken house to stimulate winter laying and you would need to invest in an incubator, but the increase in return would be worth it. Do be sure of your market, take orders before you set the eggs. You do not want to be stuck with too many chicks to rear yourself as they can prove expensive to feed.

Another improvement on this is to use a self-sexing cross or sex linkage. Here a cock of a dark breed is used on hens of a light breed producing chicks which can be sexed by coloration on hatching (hen chicks brown and cock chicks white). An old favourite cross is Rhode Island Red cock on Light Sussex hens which produces a dual purpose bird with hens that are reasonable layers and cocks that can be fattened for table use.

Low input philosophy

Now let us turn our attention from maximising profits to helping cash flow. Most modern livestock units rely heavily on bought-in food. This food is produced to suit the nutritional requirements of a particular animal and even caters for various seasons of the year, including lack of grazing in winter. Bought-in food is also reasonably expensive, costing between about £4.50 and £6.30 per 25 kg sack. If an animal can be kept on an extensive system it is going to find more food than it would if kept intensively; therefore a pig allowed to range in a paddock or orchard will provide some of her own food by grazing and foraging, but do make sure that her nose is well ringed otherwise she will root everything up and produce a landscape akin to that after trench warfare.

To take this a step further, grow a fodder crop especially for the animal. This starts to close the agricultural circle; the land provides food for the animals and the animal provides dung to fertilise the soil. A closed circle should always be aimed for because, by minimising waste, savings can be made. There is a problem here and that is time. The time involved producing a patch of roots to feed your pig(s) may cost more in time than her 5lb per day of barley meal would cost in hard cash. However, if cash is scarce and time is plentiful it is a good investment. The usual crop for fodder is turnips or fodder beet, but kale can be grown. Apparently comfrey is used in some Third World countries and has a high protein content; this is the type of thing that needs further investigation.

Low input management is really turning the clock back to traditional farming techniques where the mixed farm was virtually a closed unit; it was truly 'self-sufficient'. We can attempt to emulate this state, but only if we are willing to forgo virtually all the comforts of modern life. This would mean disposing of all mains services and motorised transport, money would become very difficult to come by and what little we had would go on rates (poll tax) or similar. This is not a practical way of life; it has too much of the 'golden age' about it, something that never was. Use the low input philosophy to prevent your hard-earned money being drained into the pockets of the feed merchants, but do not let it take up so much time that it stops you earning money.

Gardening comes into the low input scheme, but grow those crops that bear well and store well. In the summer green vegetables are freely available, but in the winter they can save money on the housekeeping bill. If you have to break new ground for your garden or have to renovate one that has been neglected, a couple of pigs prove very efficient at rooting everything out, and remove a lot of perennial weed roots. But be prepared to do some serious ground levelling when they have finished; also worm them regularly.

Finally, remember that the purpose of low input is to save money. Do not go out and buy an assortment of farm implements to cultivate a small patch of roots. This may sound obvious, but it is easy to persuade yourself that you need a tractor, plough, harrow, rotivator etc. Unless you plan to do contract work for other people it will not make economical sense. Far better, as you get to know people, to try to share equipment or swop labour, but remember an hour's manual work does not pay for an hour's tractor time. When it comes to ploughing or ditching it is well worth doing a day on hay making to get a couple of hours' machine time because in that couple of hours you will get more done than could be achieved in a week with hand tools and the finished result will be far superior.

General care of livestock

Before you start to fill your land and buildings with animals it is a good idea to know what to expect in the way of feed consumed and space required per head and how much time they will require. Many specialist books are available providing information on food intake and conversion rates; the same books give a good idea of stocking density although, if your accommodation is not purpose-built, you may have to modify this slightly. The one thing that is very difficult to obtain from these books is how much time you will have to spend tending your stock. It is impossible to take labour figures for a large purpose-built unit and simply scale them down to smallholding size; you will always find more time is needed. The only way to find out for sure is by talking to other people and helping them with their stock. While you are doing this you may see a better way of doing something which you can implement at the start for yourself.

Housed livestock should always be kept in dry, sanitary conditions. It is pointless to spend good money on an animal only to let it die for lack of care; if it does not actually die, it will not do as well if neglected. Simple guidelines for animal housing, if followed, will improve the conditions and ease of maintenance. The pen floors should be concrete and the walls rendered smooth, allowing for ease of cleaning and disinfection periodically. The floor should have a run to aid drainage and there should be somewhere for the drainage to go. The building should not be stuffy and airless because this can lead to respiratory problems, but howling gales should not blow through them at ground level. A movement of air above the animals should be aimed at and inlets/outlets should be baffled to stop strong winds. Pigs do enjoy more warmth than most other stock, but they do have less coat. Good straw for bedding is used for most animals, but when litter is required as for housed poultry soft wood shavings can be very useful, if not too dusty.

The same specialist books will include sections on various ailments afflicting the stock. There is a temptation just to skim this section, feel depressed at the numerous ways animals can die on you and pass on to something less depressing. Unfortunately, you must read this section and know the symptoms, outcome and any possible cure for at least the more common illnesses. Many things are easily cured by using the appropriate medicine at an early stage. Calves are notorious for dying quickly. If a calf starts scouring and is left untreated it can be dead in a day. If you had (and we do not suggest that you do) spent £250 to £300 to buy the calf it would be a disaster. Bucket reared calves are hard work, can give trouble if the nutrition is not just right, and never do as well as a calf reared on the cow. Cattle in general are better left to the larger farmer. If you really want a house cow, a pedigree Jersey kept as a 'pet' is possibly the best way to go about it.

If you have an animal that does not appear to be well and you do not recognise the symptoms, a phone call to your local vet will probably clarify the situation. However, once you know the problem and the cure you should be willing to administer the medication yourself. Most animals need injections at some time and you should watch them being given by a proficient person before you try; but you really do need to be able to administer them yourself. The injections most often given are penicillin for a wide range of infections and an eight in one for ewes prior to lambing; this protects the lambs from various diseases early in their lives.

Other routine work is the regular worming of any stock; even free range chickens should be wormed as the thriftiness of stock helps your profits as well as the animals' well-being. If you have sheep, know about foot rot and how to deal with it. It is most unpleasant for the sheep and affects their ability to get around and graze, leading to a drop in overall condition. The best way to deal with it is to trim and spray or dip the hooves regularly once a fortnight and, if possible, move the stock on to different pasture regularly. There is an old saying that sheep should not hear the church bell ring twice on the same pasture, indicating that they were moved weekly: presumably an early attempt to control various parasites which become a problem when land is overgrazed.

If stock is kept well and given that little extra care it will repay you by being productive and minimising losses. This is where a small unit can score over a larger one, but you need as much skill as possible as soon as possible. Vets' bills are expensive, but a dead animal even more so. If you are in doubt at any time call the vet. There is a series of illustrated books called *The TV Vet* covering a wide variety of livestock; these are a very useful addition to the bookshelf.

Whatever treatment an animal requires the most important thing is to administer it promptly. Do not put it off 'to see if it gets better by itself'; the chances are that the animal will not, and may possibly die instead.

Movement of livestock

There are rules and regulations concerning the movement of animals. Many of these are common sense and are aimed at animal welfare. Others involve paperwork and are aimed at making animals traceable. Both are important and should be complied with.

When transporting animals they should be in a suitable vehicle with sufficient ventilation. They should not be overcrowded as if one lies down or falls down it could be unable to regain its feet and could be trampled. A small trailer with high wooden sides is quite satisfactory for stock if a tarpaulin can be lashed over the top. With the tarpaulin removed it is useful for moving timber, galvanised sheets and other building materials.

The paperwork involved with animal movement is the movement book and pig movement licences. It is essential to record all stock movements in your movement book as it is open to Ministry inspection at any time and you are required by law to keep it up to date. The pig movement licences are issued by the local Estates Office and are required for any pig moved other than direct to the abattoir for slaughter when the pig should be marked with a red cross on its back.

This paperwork is dealt with more fully in Chapter 6 with the sheep dipping papers.

If you plan normally to move only one or two animals at a time, but occasionally want to move more, a litter of porkers to be slaughtered, for example, it is pointless to keep a large trailer for these occasions. It is better to have your small trailer for your frequent jobs of fetching the odd bag of food or taking a sow to the boar and call in a contractor (or neighbour) to do the occasional heavy haulage. Heavy is the operative word. A litter of

10 porkers averaging 110lb live weight give half a tonne; add to that the trailer and you need a reasonable car to pull the load. It is better to pay someone twice a year than to have a large trailer idle all year and keep a suitable tow vehicle when a smaller one would be cheaper to run on a day-to-day basis. The only way to justify this would be to offer an animal transportation service yourself, but before doing this check on the availability of similar services and what they charge per head. There are usually people at the markets willing to transport any stock that you have purchased. They have often brought their own or someone else's in for sale and are looking for a return load. If you consider doing this kind of haulage it will be necessary to attend the markets regularly so that people know they can rely on you to be there.

Purchase of livestock

Once you have read all the books and questioned anyone with any knowledge willing to give you an answer you will have to purchase your stock. When you start viewing stock it is wise to take with you an experienced person whom you know and trust. There is usually someone who is willing to show 'a new boy' what to look for and various good and bad points. Where you look depends on what you wish to buy. If you are looking for pedigree stock you will have to attend specialised auctions run by breed societies or a dispersal sale when a herd or flock is being broken up for some reason.

If this is your plan you will probably have a good knowledge of the breed yourself and know what to look for; otherwise it may prove difficult to find someone with the specialised knowledge needed to pass judgement on the quality of the stock under the hammer.

On the other hand, if you are buying more mundane stock (less expensive stock!) with which to develop your skills and increase your knowledge of animal husbandry, many local markets will be able to supply your needs. You will also be able to find a number of people who will sell you stock privately. Many of these people will be perfectly genuine, but ask yourself why these animals are not for sale in an auction where they should attain their best value. The answer to this may be that the vendor does not want to transport them to the market and then have to pay auctioneers' fees or they may not be very good quality and so would not sell well.

Try to build up a picture of what fetches the best price and how much you will have to pay for what you want. Then if a genuine good buy presents itself you will recognise it and act quickly enough to secure it before someone else does. In general there are very few bargains when buying livestock. There are good buys and bad buys and the bad buys usually occur when someone is trying to be clever and get a 'real bargain'. Buying a flock of ewe lambs at butcher's prices can seem good sense when looking at the prices of one- or two-year-old breeding ewes, but they will only lamb at possibly 70 or 80 per cent that first year, they will lamb late and you could have problems with them because it is their first time (especially if it is yours as well). It is better to buy fewer animals of better quality than to waste your resources on inferior stock. This applies to your sire as well; it is impossible to get good offspring from a bad sire. Remember the ram is half your flock and this is true for all animals. If you only have a few sheep it may be possible to use the services of a neighbour's ram, saving you that expense.

Different breeds of a particular animal are suited to different environments. Some can tolerate cold weather, others need to be inwintered; some can exist on very poor grazing, others need good quality grass to flourish. The suitability of the stock is arguably more important than their absolute quality because a poor example of a breed kept under the right conditions will thrive better than a good specimen kept under adverse conditions. Do try to keep the best you can afford and manage them as well as possible. This way maximum enjoyment is obtained as well as maximum profit. A quality product always realises the best price no matter what it is.

Stocking levels

This is something that nobody can advise you on without knowing the land being stocked and the type of management proposed. The *Farm Management Pocket Book* (John Nix, Wye College) gives stocking rates of between three and nearly five ewes per acre. This is a good range to work with, but as most smallholdings are on marginal land a maximum of four would be better, unless you are sure about the stock-carrying capacity of your land. Talk to local people who may have experience of it in the past; however, you must remember that the land condition could have changed for better or worse.

If you plan to have a house cow, allow about an acre for her grazing, again more or less depending on pasture quality; also expect to have her inside for five months starting some time in November. The general rule should be to understock until you know the land because, if the land is poached in winter, the grass will not come so well in spring when you need it.

You can increase the productivity of your land by using artificial fertilisers. They are relatively expensive and recently their excessive use has received a lot of bad press. However, used in moderation they can increase the food available from your land. It is important to get the nutrient balance correct and for this reason there is a selection of fertilisers available with different nitrogen, phosphorus and potassium ratios (NPK). The acidity of the soil also affects plant growth and nitrogen uptake and this is controlled by the application of lime. The only way to know how much of what is required to bring your land into the optimum growing state is to have it analysed. This can be arranged through the fertiliser supplier, and is a free service to large users. You may get the service free directly or if not ask a local farmer if you could slip a sample bag of your soil in with his. These samples are taken in a W shape across the field, avoiding corners and edges as these are not representative. The actual soil for analysis should be taken from two inches below the surface which is where the grasses draw their nutrients.

If for any reason you do not like the idea of using artificials, which goes against low input philosophy and organic growing, you will have to use dung. This may be deposited as the animals graze or collected from in wintering sheds and spread as soon as the land is in a fit state to take the dung spreader with minimal damage.

In many cases the stock-carrying capacity of the land is not controlled by the food available, but by its condition in wet weather. If the land turns to mud as soon as it rains heavily all the dung or fertiliser in the world will not help. The ditches must be clear and the drains (if present) must be flowing, otherwise your money will literally be going down the drain.

Resale value of stock

You always find that when you want to buy something there is a vendor ready to take your money, but when you want to sell something, nobody gets killed in the rush. This may exaggerate the situation slightly, but it is always easier to spend money than earn it.

For this reason be sure that anything you buy comes at no more than a fair market price. This applies to animals and hardware alike. The only difference is, if you buy livestock and do nothing with it, the longer you keep it the more it owes you because of the feed bill; but if you buy hardware (deadstock) keep it as long as you like - it never eats anything.

Nobody makes the right decision every time so there is a chance that something bought will have to be resold; if this is done at a loss your working capital is eroded. The worst possible scenario is that you need cash quickly for some unforeseen reason. Your stock may be the only source available (other than borrowed money) and for that reason should be worth a similar sum to that paid and be easily saleable. This is where fringe farming ventures are risky; it could prove difficult to disperse a breeding herd (flock?) of esgargots or crayfish quickly because of the lack of buyers.

Chapter 5

What You Came For

Smallholding and having a job

Smallholding enterprises tend to consist of a patchwork of widely differing things, some agricultural and some not. As time passes these will sort themselves into various categories: those which make money and are worth doing; those which do not make money and therefore are not worth doing unless you really enjoy them; and those that lose money and must go. If something falls into the middle category and just breaks even, you must not spend much time on it or else it will be depriving a more profitable enterprise of your effort. Possibly the best way to start any project would be as a hobby which could be developed into a paying concern over a period of time. To do this, though, a regular income would be needed which should not be drawn from the finite source of your capital. In short, you would have to get a job.

Full-time employment while smallholding

Few people take up smallholding with the intention of ever working full time for an employer again. The whole concept of smallholding is one of freedom of choice, to do what you want, when you want and generally be your own boss. This is excellent if it works out like that, but many people find that, after their honeymoon with smallholding, money starts to get tight. As described earlier their capital of goods and money brought with them starts to dwindle and they do one of three things. They give up and go back to their old life or at least their old way of making money. They carry on regardless hoping things will get better, spending all their capital and then all their overdraft until they are forced to sell up. They then get a job, to pay their debts. The trouble is this is the wrong time to get a job. The smallholding will already have been developed into a full-time occupation which means something suffers, either the stock or you. It would have been better if the job had come before the stock. Then the capital would have been preserved and you would not have overstretched what you could do in the time available, and you would have avoided any debts which might have been incurred along the way.

Although rural areas tend to suffer from high unemployment there is usually something to be found, but the chances of finding a well-paid job in your field of expertise are minimal. You have to take the best that is available at the time and keep an eye open for something else to crop up. You could reasonably hope for £2 up to £2.50 per hour, although this will vary with area and what you are willing to do.

As you can see this gives an income of around £100 per week and takes all day, five days a week. This leaves early morning, evening and weekends to be a smallholder. This is not easy and in the short days of winter can become positively unpleasant, unless it can be arranged that one partner works full time while the other runs the smallholding. You probably feel that this is not what smallholding should be about, but if viewed as a short-term measure to overcome a specific shortfall in funds it is better than going broke.

Part-time employment

It may be possible to obtain part-time work in the area allowing you more time to run the smallholding. Theoretically, this should be better, but if evening work, ie bar work is done it has an effect similar to working full time. You work all day on the smallholding, then work all night and for how much? Sadly, part-time work is often lower paid than full time and you could only expect between £1.50 and £2 per hour. If you worked four hours per night five nights per week for £1.80 per hour you would receive £36, a useful sum, but no one wants to do hours like that indefinitely on top of a day's manual work.

The best way to combine smallholding and work is by jobbing, taking casual work whenever you can find it. This may be anything from agricultural labouring or logging to painting and decorating; basically anything you are capable of doing or learning to do. Again, it will not be well paid, but it will integrate better with the projects that you are setting up at home. If you use the income from jobbing to finance the setting up of a non-agricultural business of your own, this will enable you to increase your earning potential in easy steps. You can then move towards the ultimate aim of working from the smallholding full-time on a variety of ventures both agricultural and other, without jeopardising your financial position.

The way you approach the problem of setting up and financing your smallholding until it is self-supporting will depend on personal preference and financial position. However, we urge you to be prudent when spending capital 'just living' as it is the correct use of this money that will determine the success of your venture.

The venture

There seems little to be made trading in one area. To make money commodities have to be moved from one area to another. This makes sense when you consider that environmental and market forces affect everyone within an area. If, for example, you decide to deal in second-hand cars with a few repairs as well, there will be a going rate for a mechanic's time and also a going rate for second-hand cars in the area. If you could purchase the car from a large auction where the general price was low but in an area where mechanics' time is priced more highly, your potential profit would have increased.

Smallholding ventures

These can be divided into agricultural and non-agricultural ventures. We do not say that this list is complete or that the arguments for and against them are definitive, but they are designed to be thought provoking.

Agricultural

Sheep
Meat lamb production is the most straightforward system to aim for. Each ewe will earn you approximately £35 per annum. This figure is true for one sheep or 100 because they are kept extensively and therefore feed costs are minimal and confined to the winter quarter. The ewes will need dipping once or twice a year, depending on the rules in force at the time. They will also need dosing against parasitic worm and liver fluke infestation a number of times during the year. However, except during the lambing season they only need looking at daily to ensure one is not stuck in a fence or has suffered some other mishap. This means they take up very little time on a daily basis.

Pedigree sheep production. If this course is taken the running costs are similar, but your initial stock purchase will be more costly. The lambs produced will be more valuable but they will need keeping longer to realise their full value. If you are breeding for ram production, you could sell ram lambs in the autumn of their first year or keep them on until they are yearlings when they will be more valuable. The problems here are ensuring that you have the right breed, presenting your product well and getting a reputation for good stock. A point worth making is that only one ram is required for every 40 or 50 ewes. This could lead one to consider the production of breeding ewes. These ewes for meat lamb production are often cross breeds between a White Face ewe and either a Border Leicester or Blue Faced Leicester ram. The offspring from this cross are referred to as Mules or Welsh half bred depending on the parents. Although the ewes are very good mothers and are sought after, especially as shearlings, the ram lambs have to go for meat and are sometimes slow to fatten. The returns per ewe would be higher for either pedigree ram or cross bred ewe production than meat lamb production, if everything went according to plan. However, the possibility of something going wrong is greater if your experience with sheep is limited. It would be best to start with a hardy and forgiving breed such as Speckle Face or White Face for your ewes and use a good tup (ram) such as Suffolk or Texel on them to produce a butcher's lamb. This will give you experience of managing a flock and produce an easily marketable product at the end. Most abattoirs will buy your lambs from you but take advice about best weights and grades to kill.

Calves

Calves are usually sold at around ten days to two weeks old; at this age they are still very delicate creatures and need careful attention. The prices at a calf market will be determined mainly by what the dealers are buying - often calves for export. The bottom line is there is little to be made calf rearing in a small way, but plenty to be lost if one dies. Cross bred Channel Island calves are generally cheaper than Friesian crosses as they do not beef so well. If you wish to try bucket rearing a calf from this early age, try to get one direct from a farm. This will avoid the stress of market (for the calf) and also the chance of it being exposed to infection from other calves. The scours is the main problem encountered with calves, which can be brought on by incorrect feeding, chill, stress or infection. It can be fatal and if the calf shows signs of being off colour and is passing very watery or white motions consult a vet immediately. The milk substitute for calf feeding is available from the feed merchants and one bag (25kg) should wean a calf.

If your land permits, a suckler cow would be another way of calf rearing. Here the cow looks after the calf and feeds it regularly. Unless the grass is plentiful you may need to feed the cow, especially if she is double suckling, ie has two calves to feed. The problems occur when trying to persuade a cow to take a calf that is not her own. Some will not complain, others will refuse to have anything to do with it. The point to note is that calves reared on the bucket never grow as well as those reared on the cow.

Chickens

Chickens really are part of smallholding. The sale of eggs is a useful, if small, income and any cockerels reared can be fattened for the table. The production of eggs and chicken meat is now a very intensive branch of farming and as such has little bearing on smallholding. However, free range eggs usually find a ready market at the right price. If you can arrange that your chickens lay during the dark winter days by giving them extra light you will find these eggs are especially saleable as many fowls will be off lay until the spring. A simple time switch and a 40 watt bulb will be satisfactory; the day length can be slowly extended to about 14 to 17 hours. If the extra light is given in the morning they will roost naturally at nightfall. If you only want eggs for yourselves and a few to sell to help cover the feed cost, no more than one or two dozen chickens would be needed. Consider 20 chickens, each laying five eggs per week (in season). This will give you eight dozen eggs to use or sell. Many more than that and your friends may become eggbound and a retail outlet would need to be found. The breed that you keep will depend on what you want. If you go for egg production alone and do not want any poultry for the pot, one of the modern laying breeds would be best. These are light birds originally bred from Leghorns and convert food to eggs very efficiently. If you want a dual purpose breed the Rhode Island Red or Light Sussex would be a better choice (or a cross between them). Whatever breed your main flock consists of, it is always useful to have a couple of bantam hens around the yard. Bantams are very self-reliant little birds and are excellent sitters and mothers. They rarely desert a sitting of eggs once sat and guard the chicks once hatched, teaching them to fend for themselves.

Ducks

If you want eggs for your own consumption, we favour ducks. A

Khaki Campbell duck can lay 300 eggs in a year and seems to find far more food around the fields than a chicken. The market for duck eggs is limited; people who like them do so a lot while others will not even try them. Their shells are more porous than a hen's egg and the duck is not so fussy about where she lays it. This can lead to contamination and, if not properly cooked, food poisoning. However, we enjoy a lightly fried duck egg with bacon and have never suffered any ill-effects and find the flavour superior to even free range hens' eggs.

If you wish to keep ducks for meat the Aylesbury is the breed to go for. These white ducks fatten rapidly, reaching a killing weight in around eight weeks if correctly fed. If a round of hotels and restaurants were established there could be a profitable venture here, but see a specialist book for further details regarding fattening rations, housing etc.

Geese

Once the traditional Christmas bird, the goose has now been superseded by the turkey. This has happened because the goose does not adapt readily to intensive rearing techniques. A trio of geese can provide a useful seasonal income during the Christmas period, but they do eat a lot of grass. Like sheep they feed off the land rather than out of the bag, so if they are not displacing some more profitable eater of grass they can be regarded as virtually free meat.

If you intend to keep some geese remember they are large, powerful and sometimes aggressive birds. The ganders can become particularly aggressive while the goose is on a sitting of eggs and should be given a wide berth at this time, but at any time geese are not a good thing to let children look after. When the goose starts laying in the spring the first five or six eggs should be removed and used for cooking as they may not be fertile; they make an excellent omelette. Geese are often accused of being over fat for modern tastes, but if they are not fed with excessive grain to try to attain a heavier weight for killing they produce a moist meat without excess fat and with more flavour than chicken or turkey.

Turkeys

Again a good Christmas trade, but the factory farmed birds come at such a low price per pound the smallholder has to aim at the 'fresh bird' market. The chicks are quite expensive and can be rather delicate for the first two months of life. If turkeys appeal to you read the books available and follow their advice as closely as you can adapt it to your own situation.

Rabbits

Rabbits are an all-or-nothing venture. You either keep one or two does for stocking your own freezer or you keep hundreds and sell them to a meat packer. If you are keeping them for your own consumption the does can be regarded as pets and the offspring as food. Keeping a buck for so little use seems a waste; try to team up with other like-minded people and share a buck.

If you intend to be a rabbit farmer some considerable capital expenditure in cages, feeders and drinkers can be expected. Check on the situation of packing stations or dealers in the area you have chosen. Rabbits are normally sold live and are either collected direct from the farm or from a central collection point such as the local market. The profit per head is small and large numbers are required to make a significant impact on the bills.

Quail

Another intensive venture. These little birds are kept both for meat and eggs. They are a high value product and if you can find suitable outlets for them or their eggs they could prove very profitable. The establishments requiring this type of produce are high-class restaurants and delicatessens. The ethnic community also holds them in high regard. They are kept indoors under a controlled environment and are fed a high protein ration. The stock that you start with is all-important and you should buy this from a reputable breeder who will give you the dos and don'ts for successful quail rearing. Birds start laying at 42 days old or are ready for slaughter at the same age.

Guinea fowl

In some ways these birds could be considered as similar to quail. Their marketing would be similar and they would appeal to the same type of establishments. They tend to be rather wild and like freedom; they are also very noisy and this trait can become annoying.

Poultry in general

If you are keeping any poultry for meat it needs to be sold 'oven ready' to attain its full potential. To do this it must be presented in good condition. It needs to be well plucked, no rips in the skin and cleanly drawn - in other words a professional-looking job. This requires considerable personal skill and time or a plucking machine of some sort or another. If you are considering commercial poultry keeping for meat, a plucker is a must.

Fish

If your property has natural or man-made ponds on it you may consider fish farming. Trout farming has been practised for many years and, if the water is large enough and well situated, can be combined with sport fishing and has proved to be a profitable venture. Value can be added to this kind of venture by producing such things as smoked trout or trout pâté which would be sold to similar outlets as quail. Have an analysis done on the water before stocking to ensure it is suitable for trout. The local water authority will do the analysis for you or tell you where you can get it done.

Another angle (no pun intended) on fish farming is the rearing of fancy goldfish. Publications on this subject show it to be profitable but requiring substantial capital input. With existing water such as disused water cress beds and an inventive mind this capital could be cut to a manageable level and with suitable outlets an interesting alternative farming enterprise could be developed.

Crayfish

Another delicacy, which could prove profitable if the correct market is found. Crayfish can be kept in any healthy lake if the water conditions are suitable. This requires a PH of 6-7, ie neutral to very slightly acid.

Once a breeding population had been established the crayfish are harvested by trapping. Stock crays, traps and information can be obtained from the British Crayfish Marketing Association.

Pigs

The pig has always been the smallholder's friend. They do not need excessively large amounts of room, but do not like being cramped either. The best way to start with pigs is to buy a couple of weaners. These should be around 50lb at eight weeks old and if fed correctly should be ready for killing at 16 weeks and weigh 110-120lbs. This will give a carcase weight of around 80lbs which is ideal for most family meals. If a carcase weight of 140 lbs is reached, the chops are a real meal weighing over 1lb each and the joints weigh 8-9lb, which is fine if you have a large family - otherwise you soon get tired of cold pork and pork curry.

Pigs appreciate vegetables and garden waste, but there are strict legal obligations for those who wish to swill feed. These rules include boiling times and segregation of cooked and uncooked swill. This makes swill feeding an unrealistic proposition for the small pig keeper. A better proposition is the use of dairy by-products and barley meal, otherwise the food has to come out of a bag. When fattening two porkers this is not too bad, but if you fatten a whole litter of ten you find the feed bills soon pile up.

If you wish to keep a sow the lop-eared breeds such as the Welsh are generally more placid than the prick-eared like the Large White. It is important to be 'friends' with your sow so that she trusts you during farrowing. When farrowing the sow goes into a trance-like state and shakes. Be cautious because, if startled, she may snap and a pig bite is very nasty. As the piglets are born you should snap the cord and place them in a box of soft straw under a heat lamp to dry off. Once they have all been born and the after-birth has come away they can be returned to the sow to suckle. Be ready to untangle a piglet from the after-birth as there is often one there. Pigs, like all other farm animals, should be wormed regularly to keep them thrifty, ensuring good use of food. If you sell pork privately you can expect to get 20p per pound more than if sold through the abattoir. On an 80lb carcase this would be £16 per porker more. At present pork sold privately by the half pig is fetching 65p per pound; this includes half the head and two trotters.

Goats

As a financial proposition goats are fine if you go all the way with goats' milk products and the necessary marketing. Otherwise a goat can be kept as a household pet which produces something useful, ie milk and meat (kid meat is excellent). Any thoughts of money saved by keeping one or two goats tend to be outweighed by their feeding costs and the time spent on them. The stories of goats eating anything are not true – they are very fussy eaters. The nearest they come to it is that they will eat anything you do not want them to or anything that is bad for them.

Bees

Keep bees only if you really want to. It is a fascinating hobby and in a good year can be financially rewarding. However, the changes in our climate over the last decade have made these good years few and far between.

Dogs

The present high price of pedigree puppies has caused people whose only interest is financial gain to enter the dog breeding market. This is sometimes at the expense of the bitches' well-being and for this reason tighter controls are being introduced to regulate dog breeding.

However, if you are going to have a pet dog by all means have a pedigree bitch and take the occasional litter of puppies. This does not add to the problems of over production of puppies and will give you a cash bonus which will cover your pets' food bill. Be sure that all your paperwork is in order and that your bitch is registered with the Kennel Club.

Do not allow cross-bred or mongrel puppies to be produced. They will only add to an already large problem; if you doubt this go to any animal sanctuary and see for yourself. Keep your bitch safe during her heat and make sure she only comes into contact with the dog of your choice. If you have a bitch that you will never want to breed from, have her spayed.

Fruit, vegetable and nursery work

If your location allows the production of fruit and vegetables and has somewhere you can catch a passing trade, a good business can be built up. This ties in well with egg sales. Remember a satisfied customer will tell others and come again; a dissatisfied customer will still tell others but will not come again. Do not sell anything that is not up to standard but use it yourself. If your holding is near an area of reasonably high population density you may consider a pick-your-own enterprise combined with a small nursery selling bedding plants, house plants etc. For this you will need good access and parking and also be willing to wait for customers which may prove very time-consuming with little reward at certain times of the year.

Excess land

This is not a problem you are likely to have for very long but, if during your first year you do not have the stock to eat the grass, you can still make money from it. You can tack the land out. This means allowing someone else to graze their animals on your land and to pay you for the privilege. How much you get depends on the quality of the grazing and how much is available in the district. The figure would be in the region of £50 per acre for a period of about six months.

If you do not want other people and their animals on your land you can sell the grass standing to a contractor who will then come to cut and bale for either hay or silage. By doing this you take no risks and put all the responsibility on to the contractor.

It would be potentially more profitable and definitely more risky to have the grass cut and baled at your own expense and then sell it yourself.

There are many other ventures which I have not covered such as fancy birds, the Japanese wood mushroom which grows on oak stumps, snail farming, worm farming or even maggot farming. The possibilities are endless but the outlets must be identified before you spend money on setting up the venture.

Non-agricultural

These are very difficult to list as the number of possible ventures is so great and what you do depends completely on what you can do and what you want to do.

Car sale and repairs

This is something that many people try but you can tie up a lot of money and, unless you are well aware of the different problems afflicting different models, you can soon lose money as well. The best way is to have a vehicle that is used by the family for daily transport and is permanently for sale, with another one ready to take its place when it is sold. As you are using the car daily this gives you and the potential buyer confidence in its reliability. Also, any repairs you do tend to be done well because if it lets anybody down in the first instance it will be you. The best place to buy is the large auctions; the best place to sell is the local newspaper or even the village shop window. Remember your mark-up will be a percentage, so more money is to be made on more expensive (newer) cars, and these cars should be easier to repair. After a while you will find out what type of car, at what price, sells best in your area. Then you can start to specialise and look for them at the auctions.

Carpentry

Depending on your level of skill you will be able to do a number of different things. If you are particularly good at carpentry you will have little trouble finding work and you will know where to look for it. If you are like most people and 'get by' with carpentry, less detailed work such as chicken houses, dog kennels or possibly rustic garden furniture is for you. Wood is expensive to buy new, but there are plenty of people specialising in second-hand timber. If second-hand timber is used it should be de-nailed and sanded to make it look presentable at least. When the professionals are selling a 12-bird chicken house for around £150 there is plenty of scope for undercutting them. Where garden furniture is concerned try the local pubs. An order to make half a dozen or so tables and benches for a beer garden would be an excellent start to your new business.

Crafts

The word craft is used to describe anything originally made by hand which is now made by machine or has been replaced by a similar machine-made item. The only reason an item is now made by machine instead of by hand is that it is much cheaper. People keep the traditional crafts alive because they enjoy doing them, but very few actually make a living from making craft items. The only way to make a living from crafts is to supply the raw materials for others to pursue the craft of their choice in an area where they are not freely available. It is easy for a smallholder with a few hazel bushes to go out and cut some walking stick blanks, but for the town dweller obtaining these could pose a real problem.

The same applies to the raw materials for natural dyeing. If you wish to use some nettles for dyeing wool they probably grow in a corner of your field; again the urban dweller would have difficulty locating them without a drive into the country. If you are particularly skilled at some art or craft try to use it to its best advantage. Look for the commercial side; do not feel there is anything wrong in using your gift to make money - it is all a means to an end and that is financing your life on the smallholding.

Agricultural vehicle maintenance

This is a far more down-to-earth topic than the last one. If you are of a mechanical bent try to learn the basic principles of various agricultural machines. These do have a habit of breaking down at the most inopportune moment, ie when they are being used. If it is hay making or silaging thousands of pounds could be lost if the weather changes before the job is finished. A good person willing to come out at any time to do repair work will soon get a reputation, especially if he is fast and efficient in making the repair. Something that combines well with this is the capability for mobile welding. If you cannot already weld it is well worth taking classes at night school in metal inert gas (MIG) and electric arc welding and also gas if you are really keen.

To be truly mobile you need to be able to weld something in the middle of a field. Either a large gas set would be required or a Land Rover with a generator and welding set on board. Both options are expensive, but it could be a good investment and will not depreciate greatly with time, nor will it eat anything when not being used, or die!

Rare breeds
Over the past few years there has been a great upsurge of interest in rare breeds. Many people consider these breeds to be rare because they have no place in the modern agricultural system. This may be the case with some, but they are still worth preserving for the pool of genetic attributes that they represent. Once extinct these genes are gone beyond recovery. Others are now finding a place in modern farming, either pure bred or more often crossed with an 'improved' breed to give its offspring some particular characteristic. This may be hardiness or thriftiness or simply hybrid vigour from crossing two widely divergent genetic lines. Although the aim of people who keep rare breeds is to return them to a sound financial footing, this may take time. Rare breeds are certainly something to start after you have your income sorted out. If you consider them as a hobby which may pay in the end you will have great satisfaction in keeping them. The other point which may be worth considering is that these rare breeds were kept by people like ourselves in times past. The fact that they do not lend themselves to high input intensive systems does not mean they are unsuitable for the smallholder - exactly the opposite. Some may be ideally suited to the low input system where slow growth is acceptable if their feed quality is lower.

The Rare Breeds Survival Trust is the body to contact about rare breeds and they publish a magazine, *The Ark*, which will give information about various breeds and also has For Sale/Wanted advertisements. There are also specialist sales where the right stock can be purchased.

Diversity

As you set up your projects give some thought to how much time each one is going to take and how they slot together, ie time requirements at various seasons of the year. The advice to diversify into non-agricultural ventures first is sound because mainstream agriculture is in a very poor state at present and farmers with years (generations!) of experience are having a hard time. Ask yourself what makes you think you could do better than they can! The only answer lies in diversity. Try, if possible, to set up ventures that, once started, will keep going with very little effort. That does not mean no work but, if repeat orders come automatically and new trade is generated by recommendation, time and money will be saved on advertising. Another useful thing is to progress your projects logically so that something new can be sold to existing customers as well as generating new trade.

Feed bills

These can be a real problem for the smallholder and have been the downfall of many. All stock will need feeding at some time, some more than others, but it is no good cutting their ration below the maintenance level because the animals will at best do badly and at worst will die. It is better to keep a few animals and feed them well than have a lot and starve them. On more than one occasion people have had stock starving because of a cash flow problem that had arisen from their lack of knowledge concerning how much each animal needed to eat each day. A catastrophe like this can wipe out an over-zealous smallholder in the first winter, leaving him with no stock and a large overdraft. If this sounds pessimistic it is meant to. People have gone under in their first year and any warning that may stop it happening to others cannot be expressed too strongly. Watch those feed bills!

There is a lighter side to smallholding. If you have got the impression that smallholding represents a sure road to a nervous breakdown and bankruptcy I apologise for overstressing the pitfalls. There are more people running their smallholdings happily and making a satisfactory living by various means than there are those who go broke. It is just you hear more about the latter.

Chapter 6

Administration

Legal requirements

Movement book

If you keep livestock on your smallholding, you are required by law to keep a movement book. In it you record all movements of cloven-hoofed animals (pigs, cattle, sheep) on and off your smallholding. This also applies if you move them temporarily to someone else's property for dipping or mating or some other reason. Some typical headings for a movement book are:

Date
Type of livestock
Age
Quantity
Ear tag no
Movement - from and to
Comments

Although age and comments are not required by law it is a useful place to make any notes about the animal when purchased. A pre-printed book can be purchased or any non-loose leaf book may be ruled up for this purpose. You may be asked to produce your movement book by a Ministry official at any time and it is an offence not to keep it up to date.

Pig movement licence

The law requiring a movement licence for a pig has been in force for a number of years and is likely to remain indefinitely. You can obtain a licence from your local Estates Office (see telephone book under County Council). The licence is required to move a pig off your land for any reason, even if it is only temporarily to visit the boar.

The only time you do not need a movement licence is when a pig is going for slaughter, and then the animal should be marked on its back with a red cross and taken directly to the abattoir or slaughter market.

When moving a pig from your holding you must not have moved any other pigs on to the holding in the previous 21 days, nor must the person whose property the pig is going to move any pigs off his/her property for 21 days. The only exception to this rule is to take a pig for slaughter. This is to prevent disease from spreading.

Sheep dipping

Dipping over the last few years has been twice a year; however it is likely to be reduced to once. Dipping is only done twice yearly when sheep scab is a problem.

You will be sent dipping papers upon request, which you have to fill in and return to the Estates Office. The papers will ask you questions such as how many ewes, rams, lambs etc are involved. It will ask what brand of dip you are going to use, what date you will be dipping, also what time you intend to start and what time you will finish. This is to allow a Ministry official to make random checks to ensure that sheep are dipped correctly in approved dipping solutions of the correct strength.

You must use an approved dipping solution, and each sheep has to be immersed in the solution for one minute, during which time it must be *totally* immersed at least twice, ie the head under. The dipping papers will list the approved dipping solutions. The sheep must be dipped in a dip large enough for it to move around in, allowing total saturation of the fleece.

The reason for dipping is to prevent diseases such as sheep scab and fly strike.

Wool Marketing Board

It is necessary to register your flock of sheep with the Wool Marketing Board, who will give you a number. This number is marked on your woolsack(s) and tags, in which you will pack your fleeces, when you collect them. The person at the collection point is able to check you off his/her list. You will be given information about when and where to take your fleeces after shearing. It is the Wool Marketing Board who will forward to you your cheque for the fleeces. The price you get for your fleeces depends upon their quality, cleanliness and presentation.

Registration of herd

It makes no difference whether you have one house cow or a large herd of cattle, they must be registered. To do this you must contact the Animal Health Inspector at your local branch of the Ministry of Agriculture, Fisheries and Food (Animal Health

Division). You will be issued with a herd number. You must then ear-tag your cattle with this number, plus a number of your own to enable you to identify the cow in question.

Your cow/s will have to be tested annually for tuberculosis.

You are required by law to register cow/s; if you don't you could find yourself in trouble.

Insurance

Possibly the best place for anyone owning a smallholding or farm (no matter how large or small) to get insurance is the National Farmers Union (NFU). It deals with a large variety of insurances and will find a policy to fit your requirements.

Employers' liability

If you employ any staff you are legally obliged to have insurance for employers' liability. This will cover you for accidents to your employees.

Third party liability

If you produce any sort of product, particularly food, to sell as an end product, you should get a third party liability insurance. This will cover you for any claims regarding the product.

Livestock insurance

It is wise to insure your livestock. The policy should cover straying animals. Cows and sheep have a habit of breaking through fences on to the road or deciding that your neighbour's grass is far better than yours. Your policy would cover any damage they might cause. Make sure your policy also covers 'death in transit'; this covers any animal that is in transit to or from your property by road. A road accident could cause the death of a valuable animal and if you are not insured for this loss, you could lose a lot of money.

It would be wise to have the policy cover such things as your dog straying and causing an accident; if one of your animals is involved in an accident you may have a large bill to pay for damages. Have your policy also cover animals biting or kicking someone; people become quite upset about this sort of thing.

Farm vehicles

The NFU can also help with vehicle insurance, particularly for tractors. You are not allowed to drive a tractor on the highway without insurance; however, the insurance for a tractor is very reasonable and covers not only the tractor but also any attachment.

It should be remembered that tractors must be road taxed to be driven on the road. You can obtain an exemption certificate from road tax if you will only be going on to the highway to get to a field which belongs to you. This does not entitle you to travel wherever you like without tax.

Grants

Enterprise Allowance Scheme

This scheme is a good way of getting started in your business. There are, however, certain eligibility requirements for the scheme. You must have been unemployed for eight weeks and be in receipt of benefit. You must also have £1000 to invest in your business. This does not have to be pound notes in your hand; it could be in the form of a guaranteed bank overdraft. Banks are usually very good about an overdraft for this reason.

You will have to attend an 'awareness day'. This day will give you a general run-down on what to expect when starting up your own business. Two business counsellors give sound advice on where to get further advice and what not to do. The day also gives you an opportunity to discuss with a group of others what you have planned and to hear their opinions of your plans, to swap ideas and to give suggestions to each other.

After this you will be called for an interview where you will be asked to sign a contract. In the contract you undertake to keep to the conditions required for one year. One of these conditions is that you work at your business for at least 36 hours a week. They will pay you £40 per week for a year, to help you get started in your business and as an aid to keep you ticking over during those first vital months.

During this year you will be visited by a counsellor at the three-month stage to see how you are coping. Sometimes you will receive a visit after six months as well, or you may just receive a letter. If during the 12 months you fail to meet the necessary requirements, payments will cease. During this year you are entitled to two free counselling sessions. Take advantage of these sessions, especially if you come across any problems. The scheme is out to help those trying to start up a business. The counsellors are there to assist you so take all the advice you can.

Rural Development Commission and Development Agencies

These bodies exist to help regenerate industry and hence create employment in rural areas. They may be able to help with advice, premises and sometimes financial backing. The RDC operates in England only, taking over the work of the Council for Small Industries in Rural Areas (CoSIRA).

The WDA (Welsh Development Agency) operates in Wales along with the Mid Wales Development Agency.

Addresses for the above can be found in your local telephone directory.

In Scotland the relevant bodies are:

The Highlands and Islands Development Board, Bridge House, Bank Street, Inverness IV1 1QR; 0463 234171

Scottish Development Agency, 120 Bothwell Street, Glasgow G2 7JP; 041-248 2700 *and* Rosebery House, Haymarket Terrace, Edinburgh EH12 5EZ; 031-337 9595.

Employing staff

It is unlikely that you will need to employ anyone to begin with. However, you may in the future, so the following are a few tips to help you.

There are three main ways of employing staff: casual labour, self-employed worker or employee. If you cannot fill the vacancy through word of mouth recommendation, it may be necessary to advertise.

Advertising

The local Jobcentre is a good place to advertise and it is free. You can also place an advertisement in your local paper. The charge is based on the space booked (depth and column width).

Be sure to state exact details of what the job entails, such as how many hours per week to be worked, qualifications required (if any), training to be given (if any) and approximate age (if applicable). Remember to give your name, address and phone number for reply. Whether you state a wage or salary is up to you. You could either put 'pay negotiable' or state how much you are going to pay per hour, per week or per month.

Applications

Deal with any applications for jobs systematically. Read all letters of application and decide which applicants sound interesting and may be suitable for the job. Write to those who are not suitable, telling them so.

Write or phone those you wish to pursue, asking them to attend an interview; if you plan to use an application form, enclose it with this letter and if necessary also send a map showing how to get to you.

Do not forget to confirm clearly the day, the correct date, the time and the place of the interview. Ask them to bring the completed application form with them to the interview, along with their qualifications/references. Ask them to confirm the interview in writing, and give a telephone number for them to ring if there is any problem or they need to arrange another appointment.

Below are some suggested questions to be asked on the application form.

Name, address, marital status, date of birth, sex.
Education: schools attended - college/university.
Qualifications: O levels, A levels, GCSEs and others.
Hobbies.
Previous employment: Employer - name and address - dates - position and job description.
References: names and addresses.

Interviewing

When interviewing applicants, make them feel welcome but at the same time be businesslike. Make sure you have the details relating to the person you are interviewing in front of you.

Ask questions on any points that may have arisen from the application form. Get them to tell you about the jobs they have done previously. Ask the reason why they left their last job, or if they are still employed, why they wish to leave it for the job you are offering. Ask to see any qualifications or written references the applicant has brought along.

Give the applicants a description of what the job entails. State the hours that are to be worked, ie 9am - 5.30pm, with a lunch break from 12 noon to 1pm.

Discuss payment, agreeing the amount to be paid and the frequency of payment, ie weekly or monthly.

If accommodation is offered with the job, give a description of what is available. In this case, accommodation with a job is usually rent free; however, bills such as electricity are paid by the occupant. This must be made clear, also whether it is partially furnished, fully furnished or unfurnished. Another aspect which should be taken into consideration when offering accommodation is whether pets are to be permitted. A separate document should be drawn up regarding accommodation for the successful applicant.

Remember to ask applicants if they have any questions and answer them as fully as possible. After the interview, invite the applicants to look around the work place and to meet any other staff (if any). If accommodation is provided, show this also.

When you have finally made your decision, offer the position, to the successful applicant. State in your letter the date of commencement, and the time you expect to see him or her on that day. Ask the successful applicant to write and confirm acceptance. Also ask him or her to contact you if there are any queries.

When the acceptance is received, write to the unsuccessful applicants telling them the position has been filled. If your chosen applicant has decided not to take the position, you can fall back on to the next most suitable one. Any employee who works for more than 16 hours per week must be given a written contract of employment within 13 weeks. In this contract you must spell out such things as:

(a) job title
(b) date employment began
(c) payment and how paid - weekly or monthly
(d) hours to be worked
(e) provisions for holidays and payment for same
(f) sick pay
(g) amount of notice required by both parties
(h) disciplinary rules and grievance procedure.

It is worth mentioning at this point that any employee who has worked for you for more than 52 weeks must be given notice in writing. If your employee has worked for you continuously for more than four weeks he or she is entitled to one week's notice or payment in lieu. After two years' service, an employee is entitled to a week's notice for every year worked.

To acquaint yourself with further details on these matters, ask at your local Department of Employment for leaflets PL699 and PL700.

Legal aspects

There are certain legal obligations involved in employing staff. The main financial requirements are income tax and National Insurance (NI) deductions.

If for any reason you engage a self-employed person, he or she will need to present you with the necessary 715 certificates (a form of receipt which you show to the Inland Revenue as proof of the transaction). However, you do not need to deduct tax or NI. The person makes his or her own tax returns and buys a 'self-employed stamp'. If you engage a 'casual' worker, he or she does not need a 715 book. Often a casual worker is paid in cash, therefore it is up to that individual to declare his or her earnings for tax and NI.

It is more likely that whoever you employ is going to be an 'employee'. In this case you must make the necessary tax deductions and NI deductions. There is a booklet, which can be obtained from your local tax office, called 'Employer's Guide to PAYE'. The tax office will also supply you with tax deduction cards and tables; these will enable you to work out the amount of tax to be deducted from your employees.

NI deduction tables can be obtained from the Department of Social Security (DSS). NI must be deducted from anyone earning more than £43 per week (1989-90); the employer also must 'make' a payment for any employee from whom NI is deducted. This amount is subject to change each year in the Budget.

Payments to the Inland Revenue (tax) and DSS (NI) must be made monthly. For any further information on tax or NI contact your local Inland Revenue or DSS office.

You are required by law to give your employee an itemised payslip with his or her payment. It must show such things as:

(a) gross wages
(b) net wages
(c) tax deductions
(d) NI payments
(e) any other payments/deductions, ie holiday pay, sick pay.

Always remember to take great care when choosing an employee. You must pick someone who you feel is trustworthy. He or she may be dealing with expensive livestock and machinery, so be sure of experience. Try to choose someone who seems to show enthusiasm and understanding of what you are doing. It is important to choose someone you can get on with easily; good employee/employer relationships are a must; they can make or break a small business.

As a self-employed person yourself you must not forget your own tax and NI contributions. An NI contributions card can be obtained from the DSS. If you only have small earnings from your smallholding you could apply for exemption from NI payments; ask for form NI27A from the DSS.

Employers' liability insurance is a legal obligation; see page 78.

Under the Health and Safety at Work Act, employers are responsible for ensuring that their employees work in a safe and healthy environment, in reasonable comfort. The employer is also automatically responsible for business activities which adversely affect the health or safety of the general public.

Accounting

Costing

This is as important in smallholding as in any other type of business. When it is decided that a particular job needs to be done, before buying any materials or embarking on the job, a costing should be prepared. Here is a suggested list of steps to be taken:

(a) Decide what materials are required.
(b) Get at least two price quotes for each item of material.
(c) Total material amounts.
(d) Get at least two estimates for any outside labour required (ie contractors).
(e) Work out total cost.

A similar exercise should be undertaken when pricing your own products to arrive at a selling price. Remember to build in the overheads and a profit margin.

Budgeting

At various times of the year your expenditure is going to be greater than others, as is your income. If you have livestock, such as sheep and cattle, your feed bills will climb dramatically during the winter months and fall to practically nothing during the summer. Income from any lambs you have produced tends to come during the summer months.

It is advisable to make a budgeting plan. This is not always easy because things have a habit of not working out quite as you had expected. However, it may give you a rough idea of where you stand.

The winter months are the hardest time for most smallholders as little is sold during this period, but it is the time when most money is paid out, therefore allowances must be made during the summer months for those oncoming feed and bedding bills.

Remember to budget for maintenance; it is easier and cheaper to do a job when it is quite small (ie when one fence post falls over because of rot) than to leave it and possibly have to replace the wire as well.

Do not forget veterinary bills in your budgeting. When an animal is sick, it is no good putting off a visit because you cannot afford it. Non-treatment could result in the loss of a valuable animal, therefore costing you more than the vet's bill in the first place.

Cash flow

This is very important. More businesses fail through lack of ready cash than from low profits. It is no good having hundreds or thousands of pounds' worth of livestock and machinery if you have no cash to feed or maintain it. You do not need to buy a tractor if it will stand still 360 days a year, and why buy attachments if you are going to use them only once or twice a year? This is tying up money that could be earning you more money, thus keeping the cash flowing. Arrange to hire equipment that you use only rarely.

Do not stock your holding out with more livestock than you can afford to look after. Keep a smaller number, keep them healthy and well fed and they will be more productive, bringing in the money to plough back in and increase on the livestock you already have.

Bookkeeping

When you are in business, you should keep a record of all transactions. These are written up in your books, for presentation to the Inland Revenue for your annual tax assessment, when your annual accounts are prepared.

If your bookkeeping is done regularly (ie weekly) it is a lot easier and it can be done very simply. Be guided by your accountant, who can show you how to organise your accounts.

Some people will say that you do not need an accountant. However, remember that he is qualified to deal with the tax inspector – you are not. He has the time; you most definitely have not.

All you need is a double-page spread cash book. The left-hand side of the page is used for recording incomings; the right-hand side is used for outgoings. Preprinted books are available from large stationers.

Each transaction should be written on a separate line. It should record such items as the date, description of goods sold or bought, invoice number, cheque number, cost and total cost.

If you employ staff you need to keep a wages book in which to record weekly or monthly payments, and such details as date, name of employee, gross earnings, tax deductions, employee's NI contributions, employer's NI contributions and net pay.

You may find it useful to keep a petty cash book to record any small items of cash paid out daily. It helps when trying to do your books later on.

Remember to keep all receipts and to file them away safely. It is wise to ask for a receipt for all items purchased for the smallholding, including petrol. Receipts are essential if you are paying VAT.

Value added tax

You must become VAT registered if your annual business turnover is more than £23,600 (1989-90). This amount is subject to change at each Budget. All your invoices will have to show VAT details charged to customers, and give your own registration number. However, if you are registered you can claim back VAT on such items as tools, petrol, building materials etc, so be sure to ask for a VAT receipt. For more detailed information on VAT apply to your local HM Customs and Excise Office (VAT Dept); they can supply you with the necessary information leaflet and advise you on how to deal with the paperwork.

VAT is at present levied at the rate of 15 per cent. VAT returns are made to Customs and Excise quarterly. It is advisable to keep VAT separate from your other income, so that when your quarterly VAT bill arrives you have the money to pay it.

Chapter 7

Marketing

Supply and demand

The rules controlling the prices of all base market inputs are controlled by supply and demand unless an artificial external pressure alters the natural balance, such as guaranteed prices and buying goods into intervention. This tends to affect the larger farmer more than the smallholder, but anything which affects the price of feed stuffs or the price of products could mean the difference between profit and loss.

Theoretically, your entry into the market should have an effect on the price of the product. In practice, unless the market is very small, your contribution will be too small to have any real depressing effect.

Consider entering a small specialised market with a product of comparable quality to the opposition but offered at a considerably lower price. This will produce two reactions in your potential customers; first, suspicion and an assumption that the price reflects the quality; second, pleasure at having found a 'bargain' and a wish to tell others about it. Unfortunately, the second reaction only occurs if the initial suspicion is overcome. This would lead to a low sales profile at first, enlarging later as the initial resistance is overcome and confidence increases. This can lead to a situation where more orders are received than your system can cope with. If this happens, either your production techniques have to be improved to give better output, staff have to be employed, or both. Either way it will probably mean an increase in the final price of the product and hence a possible drop in orders as your price approaches that of the opposition.

Had you entered the market with a price and quality comparable to the opposition and offered a 'special introductory offer' of x per cent off, this would overcome any initial resistance to the low price, but allow you to undercut the competition and attract custom. If you feel it is unnecessary to charge your customers full list price, you can give them trade discount, bulk discount or regular order discount to keep them buying from you and no one else.

However, if you are selling a product into a large market, for example fat lambs for meat, the only control you have over your final price is when and where you sell them. Some markets tend to attract more buyers than others and hence will average a slightly higher price. You could sell your lambs direct to an abattoir where you will get a set price which you will know in advance. This is an advantage over the market where you do not know what you have made until the hammer falls, and if the lambs do not reach your reserve you take them home again, having wasted your time and money in transportation. The chances are that you will get a slightly higher price by going through a market, but once auctioneers' fees are deducted and you have accounted for the time spent there we doubt there is much to be gained either way.

Advertising

The advertising that you do depends greatly on what you have to sell. To continue with the sheep example, if you have some butcher's lambs for sale you would not get a higher price for them by advertising in the local paper than by taking them to the market or the abattoir. In fact, the only offers you would get would be lower from people who were trying to make a few pounds by doing just that. However, if you had some prime examples of a sought-after breed, maybe a promising young ram, your advertisement could increase your return by bringing it to the notice of more interested parties, although this could equally well be done by entering the animal in a ram sale.

A similar example can be given for cars. If you had repaired and MOT'd an old Ford Cortina, an advert in the local paper would be sufficient, and because this type of car is freely available all over the country advertising nationally would be an unnecessary expense which would be unlikely to draw in any extra buyers. If, however, it was a collector's car of interest and value the advert should be placed in a national magazine to reach the highest number of potential buyers.

If you are producing anything other than base products with their prices fixed by supply and demand do not underestimate the power of advertising. Even the sign on the front gate 'Fresh eggs 85p per dozen' is good value for money. It costs nothing once erected and has a direct impact on passing potential customers. You can embellish it with what you like, but the basic information (what you are selling and how much it is) must be big enough to catch the eye. If you sell a range of goods which varies during the year, it would be worth making a board with either hooks or slots to support different signs for the various products. This is much neater and more professional looking than a blackboard with the products chalked on. Not only that, but you do not have to re-write it each time it rains, and a painted sign is more readable from a car than a chalked one. This is true for all adverts; the item to be sold should appear first, followed by a description and the price. Why people advertise things without giving a price is beyond me. Someone is either willing to pay your price (or something close to it) or they are not. It is pointless having a string of phone calls from people who say 'How much is it ... Really, oh, I'll let you know.' What is more likely to happen is the person who is genuinely interested suspects a ridiculously high price and never calls at all.

If you are advertising where you are not charged by the word, you can afford to try to catch the punter's eye in other ways. There was an advert in a local shop window for 'Furry mouse traps' and had we not already got a very efficient one of our own I would have been tempted by this far more than by an advert for 'cats'. Try to catch the eye, possibly with a photograph of the goods, although this may not be a good idea if you are selling Christmas turkeys or freezer pork.

What the advert should do
It should be:

Informative, giving details of item(s) and price.
Noticeable. Make it as large as the value of the item warrants. Bold print for what it is.
Honest. Don't forget the Trade Descriptions Act. If the advert says organic, it must be organic. Simple and to the point. Don't let the punter get bored before he finds out what you are selling.
Targeted. The advert should appear where the greatest number of buyers will see it.

Targeting of adverts

Effective advertising can be expensive. Even a small advert in a local paper, if taken on a regular basis, can soon add up to a sizeable sum. For the sake of economy your advert must reach the largest number of potential customers possible; in other words it must be targeted. If you place an advert to sell a tractor in the miscellaneous column of the local paper it will be read by a great number of people, but the percentage of them looking for tractors will be low. It would be better to place it under farming or possibly even livestock because the type of person who regularly turns to that heading and reads the adverts is more likely to want a tractor than anyone reading miscellaneous. It would be better still to advertise in a specialist farming journal where people seriously interested in buying a tractor would look.

The same applies if you are offering a service. If you take an advert in the paper or put up a poster in a shop the number of people who read it may be high, but the percentage to whom it applies is probably low. In a case like this a mail drop or mail shot is the most targeted advertising you can possibly have.

If you had a system for inspecting and/or unblocking land drains, this would obviously be of interest to farmers and landowners primarily and few other people. The best way to advertise your services would be with printed sheets which gave the message in bold print so that the information was transferred as soon as it was out of the envelope. The names and addresses of farmers can be obtained from the Yellow Pages; also others like the council parks and amenities department. Many sheets will be consigned to the bin, but not before the message has been transferred; here is someone who can do this. Some will be filed for future reference and some may arrive just as the person is worrying about his drains! Had you taken an advert in a paper you would have needed a long run to have the same kind of impact as a mail shot; this makes the mail shot far more cost effective for the 'service' type advert.

The only possible improvement on the mail shot is getting in your car and doing your own direct sales. The disadvantage is that farmers are busy men who are beset by numerous sales representatives and so send them packing before they have a chance to sell their wares. 'Phone XXX to discuss your requirements' printed on your mail shot may be good enough.

This is just one example, but whatever you do will have a definable set of potential customers. Once you have decided the type of person likely to want your service or product, find their addresses and mail them. This can be followed up in about a week by a telephone call.

The object of the telephone call should be to arrange a meeting. Depending on what your product is it may be possible to offer some incentive like a free sample or a demonstration. Once you have the meeting, it is up to you to sell yourself and hence your product. The only problem here is time; while you are out selling or demonstrating you are not actually putting any money in the bank. If you can get sales by post or telephone, save personal visits for special prospective customers who could give you large or repeat orders.

Dummy adverts

If you wish to get the feel of a potential market before you decide to enter in too big a way, you could take a dummy advert. Two things come from this; first, you will be able to gauge the market from the response you get; second, you should be able to take orders for your first production. You may have just a few items to sell to any replies, but you can always say, 'I'm terribly sorry we have sold right out of this batch. If you will leave your name and number I can call you when the next lot are ready, which should be next week/month'. However, once you have taken an order you should feel duty bound to fulfil it even if it is the only one and proves to be uneconomic.

Mail order

A mail order business is an obvious choice for someone living in a rural area. Daily postal delivery and collections make servicing a huge number of clients possible. The difference between mail order and a shop is that the customers part with their money before seeing the goods. For this reason some people are suspicious of mail order firms, fearing that they will send their money and either receive inferior goods or nothing at all. This situation has come about because over the years there have been some dishonest mail order businesses set up and then closed down, taking the money with them. To protect their readers from fraud, newspapers and magazines which accept mail order advertisements operate the Mail Order Protection Scheme, and they will check your proposed advertisement - and your business — very thoroughly before accepting advertisements which invite readers to 'buy off the page'. Further information will be found in *Running Your Own Mail Order Business* (Kogan Page).

If you are going to sell by mail order your advertisement is your shop window. It is all your potential clients will see of your operation. It has to give them the confidence that if they send you money they are going to receive quality goods in return; sadly this type of reaction is not produced by a two-line minimum 20 words advert seen once every three months in *Exchange & Mart*, that gives the impression of someone in a garden shed. The customer wants to see a good advert regularly which not only promises value for money and a money-back guarantee, but inspires confidence by its size and regular appearance. To finance advertising like this will cost thousands of pounds per year, but if you do not advertise no one will know you are there. Of course, you can start with smaller adverts and work up, but if the initial ones do not produce the sales you never find out if the bigger ones would work any better.

The only way to overcome this problem is to combine your efforts with other people with a similar product (not necessarily identical) and share advertising.

Marketing co-operatives

These are groups of producers who have similar or complementary products who can increase their chances of selling their products by sharing advertising or marketing under a group name. Each producer will pay a 'membership fee' to the co-operative and for this will receive certain advantages. These may include discount on items bought in bulk by the co-operative and redistributed to its members, higher profile advertising than any individual member could afford, and the ability to sell his product through large outlets such as supermarket chains, which individual producers would have difficulty supplying.

Being part of a marketing co-operative increases the producer's marketing power and may help with such things as new markets and export. The main difference between marketing co-operatives and distributors or wholesalers is that with the former you sell direct to your customers with the assistance of the co-operative. With the other two you sell to them and they sell to the customer, obviously at a profit.

Creating markets/cold selling

This is where the marketing power of a co-operative could prove really useful, but that does not mean the individual producer cannot do it alone, although it is not easy.

Let us consider the lone producer who has an idea but no ready-made market. Initially, a potential market must be identified, researched and approached. For an operation like this to succeed the personal approach is necessary because some persuasion will be needed to start a completely new product from cold. You have to convince your potential customer that the product is just what they always wanted but could never find before. If you can use the 'as seen on television' ploy it will help. Your product has probably come from some corner of the world, so if for instance it was a food considered a delicacy in Tasmania tell the delicatessen manager so; he may not be impressed but his customers might be.

Before you go out to do your cold selling, think of every sales representative you have met, even those calling door to door selling double glazing. Try to emulate any good points and discard any bad ones; be professional and confident (without being cocky). Have some business cards printed. They are not very expensive and give a far better impression than jotting your name and number down on a scrap of paper. Have as many facts and figures regarding your product as you can. This will at least give you something to talk around while the potential customer makes up his mind. The most important thing that you must do is make it easy for them to say yes. If you have a sample ready with you, you can say something like 'I will leave this with you and pop in next week to see how you found it!' If you get a no, at least you realise it was a no hoper and can take your free sample down the road. This is a very useful ploy. 'Well, your competitor down the road is very interested' can bring some response, but don't lie outright; they might be drinking buddies.

As soon as you have completed a call write it up in a notebook. Record the names of people spoken to, what they did and did not say. Try to assess the genuine level of interest and their reaction to the price. Make a special note of time and date of call and the same for any future call arranged or promised.

If you were considering suckling pig production, for example, a possible scenario for creating a market could be an establishment catering for the reception, function and conference market. It is essential to see the decision-maker, in this case either the manager or the head chef. If you are kept waiting, chat to the receptionist. Without being rude try to find out as much as you can about the establishment. If you found they were looking for 'theme' ideas this could lead you at a point during your interview to tell the manager how successful an acquaintance of yours had been with medieval banquets, the main dish of which was, of course, suckling pig. If you ever get a break as perfect as that, you are very lucky, but it is an example of how to turn a little information to your advantage.

Finally, on the subject of personal contact sales, a man should wear a suit or at least a collar and tie. You may be a smallholder in wellies and sweater at home, but out there you are a sales representative and you are all the world sees of your operation.

Following food fads

Once you are settled in your smallholding there is a tendency to become very introverted. The world outside becomes a place of noise and confusion which is pushed into the back of your mind and replaced by the tranquillity of your own little world. Although this is very pleasant it is not to be advised if you hope to sell food into the 'fad' market.

Every few years a new food fad will come and go. It may affect the type of food eaten or the way the food is produced. There is no doubt that if you can follow these fads and detect when one is waning and another gaining favour you will be able to tailor your output to the new market. There is a negative side to food fads as well; some foods go *out* of favour suddenly and usually in a big way. At present we have just come through an egg bashing period (because of food poisoning fears). However, in a year, the chances are that just as many eggs will be used daily as there were before the scare, but in the meantime it will have created many problems in the egg producing and marketing industry. There would be little you as a small producer could do under these circumstances, other than try to allay the fears of your customers by giving them assurances of the wholesomeness of your eggs. You could use such descriptions as naturally produced, free range, from grain fed hens, organically produced - if, of course, they were true.

There are other less dramatic 'anti fads' such as the avoidance of animal fat in excess. Whereas the working man needed and wanted fat on his meat to give him energy to carry out his manual labour, now fat in quantity is unneeded, unwanted and actively avoided by many people. It would therefore be unwise to start producing pork from a traditional breed of pig renowned for its tendency to produce fatty meat or even to feed a more modern, lean breed so well that it put on excess fat.

If we consider the last decade or two the main changes in diet have initially been a swing to convenience foods of all types, but mainly those replacing traditionally produced food with something closely resembling it. There was little attention paid to what actually went into the food until later on. Then came the movement away from artificial additives and colouring; in other words a traditional food in a convenience pack. The increasing resistance to fat in the diet has been mirrored in farming by a constant search for leaner strains of animals to supply the housewife with her weekly joint. Later an awareness grew of how the food was produced. A resistance to factory farming, growth promoters, herbicides, fungicides and many of the other aspects of modern farming led to the organic food industry. Most people think organic food refers only to vegetables, but there is also a corresponding real meat movement.

Finally, people nowadays tend to eat more exotic foods than their parents did, and although many of these are imported it still opens up many opportunities for the enterprising smallholder who wishes to try something completely different. But don't rely totally on such an outsider.

The way to try and follow the movement of public opinion is either directly through the trade, ie delicatessen/specialist food shop managers, or through the media. The various food and cooking programmes shown on TV are good because they can actually start a fad; also statements from the House of Commons or the Ministry of Agriculture, Fisheries and Food (MAFF) can give information on things to come.

Creating the perfect product

From the above section it would appear that the ideal product should be naturally produced, free from unnecessary additives, not excessively fatty, convenient to use/cook and also be interesting (exotic). This is not something that could be created instantly or easily, but things that bring good returns are worth that extra time and effort to get them right.

Let us consider the pig again. If a half pig is purchased, that is exactly what you get, straight down the middle. Most people now tell you that they do not want the head and trotters or dispose of them as soon as possible. What a waste! The first things to come out of the freezer are the chops followed by a joint for a special meal, but eventually there will be some belly pork and bits of neck left; these often get wasted or are fed to the dog.

To create a good package for you as well as the housewife, you still want her to buy all the pig to avoid waste, but she must be happy with all her purchase and, it is hoped, pay a higher price for it. This package could take slightly different forms, but would aim to disguise the less desirable parts of the pig in some way. The chops would stay as they are being virtually a natural convenience food and, as they are highly priced, need little improvement. The hind leg is normally cut into two giving the fillet half leg (upper half) and the knuckle (lower half). One of these two could be boned and cut into steaks; these have the same convenience as chops without the bone and are again a high value cut. That has neatly dealt with the more expensive part of the pig. Now what about the rest? Sausages are a favourite; ensure a high meat, low fat and rusk content and this will set them apart from the mass produced item. There are many different sausage recipes available, but once you have found a good basic mix try different flavourings (garlic, curry, herbs etc). The sausages are made from the neck and belly pork, which are also the basic ingredients of pork pies. A homemade pork pie is a truly wonderful thing and bears little resemblance to the item bought in the supermarket. To make the jelly for the pork pie use a trotter which is one of the less desirable parts. To make the sausages and the pork pies would need a mincer with a sausage filling attachment and, although a domestic model would be satisfactory for practice and a trial period, an industrial model would soon become necessary once you were in production.

We are now left with the head, the hand or shoulder and the pluck (heart, liver etc). The head is the part that nobody wants and for this reason you could make a traditional brawn and include it as part of the package or, more simply, remove the meat from the head and either include it diced as 'stewing pork' or use it in the sausages. The hand also may be best used in one of the higher value items such as pies or sausages, but if cut correctly it is possible to take two or three steaks from the thin end which would be suitable for grilling. The pluck should be turned into or used as the basis for faggots or, better, various pâtés.

This uses up the pig and should double the price per pound obtained for your pork.

Each unit would contain: chops, steaks, joint, sausages, pork pie(s), diced pork/brawn, faggots, pâté.

This should be billed as good wholesome food produced in time-honoured ways without the aid of artificial additives, of low fat content and fed (if possible) organically. By doing this you could increase the profitability of pig keeping from marginal to excellent. You could take this one stage further and cure some pork and sell a package containing bacon as well as pork and its products. This is only one more step on the learning ladder and could be taken easily once the rest had been mastered.

To undertake a project like this would involve considerable planning and liaison with the public health authority. However, the public health authority will be very pleased to tell you what you need for such an enterprise and if you follow its advice you should have few problems sorting out the necessary facilities. Do not try to conduct a business like this from your kitchen as this is unsuitable.

Although the above example is very specific, the general theory could be applied to other areas. Make the product appeal in as many ways and to as many people as possible. Maximise your potential market.

Chapter 8

Safety

Many accidents happen on farms and smallholdings. Some involve machinery, but a lot don't. Children are often the victims of these accidents, usually because adequate safety precautions are not taken. Make your smallholding a safe place to be, not only for you and your family but for visitors and any tradesmen who may be on your property.

Some of the following points are perhaps obvious to most of us, but it doesn't hurt to jog the memories of others.

Poisons and medicines

Poisons must be kept in a locked shed or cupboard. Items such as rat poison, dipping solutions, disinfectants, pesticides, slug-pellets are but a few. These can not only cause trouble for a child but also animals.

Medicines for animals cover a wide range, from antiseptic sprays for new born lambs to antibiotics and syringes for their application.

Items such as navel sprays, footrot sprays, worming powders, marker sprays etc should be kept on a shelf along with your locked-away poisons. Antibiotics, calcium solutions, ointments, tablets, syringes and needles are perhaps best stored in your home, locked away like your own medicines.

Hay and straw

Haybarns are no place for children to play, however tempting. They could fall from stacked hay or straw. If they pulled a bale on top of themselves or one fell on them, it could cause serious injuries or maybe even death. Bales of hay and straw are very heavy.

Never allow anyone who is smoking near hay or straw, for obvious reasons.

When dealing with hay or straw it is advisable to wear a mask, especially with old hay as this can be dusty. If inhaled frequently it can cause chest troubles or a serious condition called 'farmer's lung'. There are special 'farmer's lung' masks that can be bought for these jobs; they have the ability to take out much finer particles than an ordinary dust mask. These are available from most agricultural suppliers/feed merchants etc.

Machinery

Machinery can be lethal if not operated correctly and safely.

Although there is a relatively low minimum age for driving a tractor, think carefully before allowing your own child to do so. Tractors are big, heavy, powerful pieces of equipment and the land on which they are driven is not always level. It is advisable to wear ear defenders when using a tractor and machinery, as prolonged exposure to the noise can cause permanent ear damage.

Store machinery safely. Attachments for tractors etc can be just as dangerous when stored, as they are when being used. Many have very sharp blades or spikes. Store them so that they cannot be fallen on to or into, and so they will not fall over. Do not allow children near them; they are not climbing frames.

Tractor tyres are extremely heavy; do not leave spares leaning against a wall where a child could pull it down on itself. Either attach it to the wall or lay it flat somewhere out of the way (but do not stack them more than a couple high).

Chainsaws should be carefully locked away when not in use. When in use, the user should wear the necessary protection, ie goggles and ear defenders at least. Children and other people should keep their distance when a chainsaw is in use, in case it should grab or the chain break.

Animals

Young children should never be left alone with large or unpredictable animals. Neither should they be given the task of feeding or cleaning out these animals.

Remember, any animal can turn nasty, particularly if it is frightened or confused.

Pigs are very heavy animals and could easily harm a child or an adult. A pig bite can be very bad, as their teeth are extremely sharp. Treat this animal with respect at all times.

Cows are big, strong animals who can do a lot of damage if they kick you. Never stand behind a cow or approach it from behind. Another point to remember about cows is that when they are bulling, and there is no bull to satisfy their needs, they will mount anything in sight, including you. So be on your guard, particularly at this time.

Sheep can be fairly docile creatures and sometimes even friendly; nevertheless they are quite strong animals, and take some holding when they need something done to them. Also, when they have lambs the ewes become very protective, seeing off anything they think is going to harm their lambs. Beware of a ram, some (like ours) are very docile, but others can be quite aggressive.

Chickens are pretty harmless, but beware of rogue cockerels. They often take a dislike to particular people, especially men, and make their lives a misery.

Geese need to be treated with great respect. They are very aggressive, particularly when there are a lot of them. Ganders are worse than geese. When they are nesting or have young they become very protective. Many people let their geese roam the farmyard, but give your postman and other visitors a thought. Remember that if someone is injured on your property, by your animals, you could be liable for damages. Being bitten by a goose is extremely painful; we speak from experience.

Remember also that animals carry many different germs and parasites. Be sure always to wash your hands thoroughly and children should be made to wash their hands when they come in from outside. Animals frequently have worms, some of which can be transferred to humans. Ringworm is a pest if your animals pick it up, as it requires diligence to get rid of it completely and it spreads very quickly.

Calves can suffer with the scours, which is particularly unpleasant. Keep children away during this time as humans can suffer from a similar complaint and it is particularly unpleasant for a child.

Sheep bring their own problems. When in lamb there is one particular complaint that can have serious consequences for humans, especially pregnant women. This condition is called 'enzootic abortion' or kebbing. It can be the reason for an aborted lamb/still birth or even death shortly after birth. Pregnant women should stay away from ewes who are giving birth; they should not assist in any births and should not handle newly born lambs. Give the lambs a wide berth for several days and ensure they are fit and healthy before going near them. The reason for these precautions is that if this disease is contracted by a pregnant woman it can cause abortion or other complications. It should be remembered also that this disease can cause illness (with flu-like symptoms) in anyone, male or female.

If an animal needs an injection, it is advisable for two people to be on hand to administer it; one to hold the animal firmly, the other to administer the injection. Care should be taken not to inject yourself accidentally. It may sound unlikely, but if the animal moves suddenly it can and does happen.

Be sure to read the manufacturer's details and instructions concerning the type of injection you are administering. Be aware of any adverse reactions that may occur if you do accidentally inject yourself. If it happens, seek medical attention straightaway, taking details of the substance with you.

Ensure that all members of your family are up to date with their tetanus immunisations. It is so easy to skin knuckles and cut fingers, but deep stab wounds or cuts are particularly nasty and anything around the smallholding is going to carry dirt and germs.

Be sure to have a comprehensive first aid kit available, carrying such items as cotton wool, plasters, bandages, gauze, sling, safety pins, sterile dressings, scissors, antiseptic ointment, antiseptic to wash wounds, crepe bandage. All items should be kept together in some kind of container, stating clearly what it is. Keep it where it can be easily seen, and ensure that all members of the family know where to find it.

Water

Many smallholdings have a well, river, stream or pond. Remember how dangerous water can be. Children always seem to be drawn to water. Make it a rule that young children are not allowed near water without adult supervision. If need be, fence it off. In any case it is wise to take preventative measures and ensure that your children can swim.

Children

Children's safety should be given careful consideration. A farmyard is no place to play. Fence off a piece of land and create a garden for them to play in, away from water, away from the farmyard, but where they can easily be seen.

Enclosed spaces

If work needs to be done in a tank or enclosed space (eg grain silo), it is advisable for two people to be present, even if the job to be done is only a 'one-man job'. Poisonous or suffocating gases could overcome the worker, causing him or her to collapse inside; for this reason a safety line should be attached to the person working inside and fixed to the outside, with one other person standing outside in case of accident, in which case he or she can haul the victim to safety and call for assistance.

Protective masks should be worn when working under these conditions if dust is likely to be present.

If, however, toxic gases are known to be present, special breathing apparatus is required and it becomes a specialist job, not one for the layman.

Fencing

This is a job that often needs to be done around the smallholding. It can result in sore and scratched hands if not tackled properly. Most stock fencing involves one or two strands of barbed wire. Barbed wire is very difficult to handle and can cause some nasty injuries, therefore always wear leather gloves when handling barbed wire. Buy yourself a proper fencing tool, for approximately £10, to make the job easier and safer. This tool grips the wire, enabling you to unwind it from a roll, holds the wire in position while you attach it to a stake with a staple, or can be used to hammer staples in; it also removes old staples. It can be used as a straining lever to position the wire in the correct place at the right tension.

When you have finished with your roll of barbed wire, put it away safely. Not only is it dangerous to leave it lying around, but it is wasting money to leave it out in all weathers, to start rusting before you have even used it. Dispose of discarded old wire sensibly. Do not leave it in an untidy pile in the fields or yard. It is dangerous for humans, and for your livestock as well. Animals have a habit of finding trouble; imagine the pain and injury which could be caused by carelessly discarded wire. You will have to free the frightened creature from the tangle, and you could have a vet's bill to pay for the repair of the damage caused to the animal.

Pesticides and herbicides

The application of pesticides and herbicides needs to be done with great care. When purchasing these items you should obtain a leaflet from the suppliers, telling you the relevant safety requirements. You are required by law to know the safety regulations relating to the chemical you are using.

Always wear gloves and a mask when using these preparations, and be responsible in their use.

Although the legal aspect applies mainly to the large farm unit, if you ever use pesticides or herbicides, no matter how small a quantity, be sure you know what is legally required of you and follow the supplier's instructions to the letter.

Electrical supplies

When electrical supplies are required out of doors, ie for outside lighting around the farmyard, be sure that all switches are waterproof.

It is wise to install waterproof switches in sheds and barns also, as they are so often damp or leaky.

Check the wiring in barns and sheds frequently, so that early detection of faulty or even chewed wires can ensure replacement before any serious damage is caused.

If you use any electrical machinery outside you must connect an earth leakage contact breaker. This cuts the power off to the machine should there be any fault causing current to flow down the earth line, such as ingress of water.

Safety of employees

If you employ ten or more people to work on your smallholding, you have to abide by many other health and safety rules and will be visited on occasions by the health and safety inspectors. If you employ anyone at all, health and safety should take a bigger role. For further information apply to your local health and safety office. However, you are unlikely, as a smallholder, to employ many people.

It must be remembered that most accidents are caused by taking short cuts. If a job is done correctly, with the proper tools and safety equipment, the chances of an accident are greatly reduced.

Chapter 9

Running a Smooth Home

It is important to establish some rules and a routine in order to keep the home running smoothly. Less time is now going to be available to do those daily tasks, such as bedmaking and washing up. If possible you must delegate some household tasks to other members of the family. Children need to feel that they are contributing to the day-to-day running of the smallholding. Give them chores to do inside the home as well as outside.

A few simple rules should be made so that every member of the family pulls his or her weight, such as, each person makes his or her own bed before setting out on the daily business, be that work or school (obviously young children are going to need help). Laundry should be placed in the appropriate place, so that you don't have to hunt for socks under the bed (it all saves you time). Children can tidy their own rooms and if old enough run a vacuum cleaner over the floor.

Do a rota for washing-up duties. All children of four years and over are capable of doing some aspect of this chore. A four-year-old is quite able to dry the majority of dishes - glassware and knives being the exception (include your best china in this category). They are also able to put dishes away.

If you have a solid fuel cooker or open fire, you will need never-ending supplies of coal, wood and sticks. An older child could be given the responsibility of keeping the wood pile stocked and bringing in the coal.

Younger children could set the table for meals and clear it after meals.

How much you give your children to do is, of course, up to you, but by contributing in some way a child will appreciate life far more. If your children are older you may be lucky enough to have help with cooking, washing, ironing and cleaning; if not there are numerous tasks a young child can do.

Children can also be given responsibilities outside the house. Younger children could collect eggs, feed chickens and pets. They could also help in the garden by picking fruit or vegetables and help with the weeding. Giving them their own small plot adds an incentive to help, and you may be surprised at how well they will do.

Older children can be given bigger responsibilities, such as feeding animals, eg pigs, cattle, dogs, sheep etc. They could also help with cleaning out living quarters, but here care must be taken that the animal is either not present at the time or is able to be handled easily by the child, eg chickens or dogs. Never leave a child with a large or vicious animal; you could be asking for trouble.

Your partner must also help with household chores if you are going to help with outside chores. It would be very unfair for either partner to expect help if it were not reciprocated.

Routine

A home can be run far more smoothly if you have a regular routine. If you all know what you have to do each day, things will be done far more quickly and chores will not be forgotten or pile up.

First, set a weekly routine for inside and outside chores. Write it down so it can be referred to. For example:

Indoors

Monday – Washing
Tuesday – Cooking
Wednesday – Ironing
Thursday – Book work
Friday – Shopping
Saturday – Cleaning
Sunday – Family day

Outdoors

Monday – Cleaning out animals
Tuesday – Farm maintenance, ie fencing
Wednesday – Check on animal feed etc
Thursday – Cleaning out storage for feed
Friday – Livestock jobs, ie foot clipping, grading etc
Saturday – Gardening
Sunday – Family day

The above are only very rough guides, to give you an idea. However, it must be remembered that these are weekly chores, especially in the indoor rota. Daily chores are many, ie washing up, making beds etc.

The same applies to the outdoor rota; animals need to be fed every day, usually twice a day. Some animals need to be cleaned out every day. You need to inspect your livestock every day. It is no good leaving this to be done on a weekly basis; a sick animal one day could be a dead one the next.

Also, if you spot a hole in your fence you should not leave it until the next 'maintenance day'. All your stock could have escaped by then; you have to do a temporary job there and then and do it properly later. Common sense is required with such rotas and they should only be used as a rough guide.

However, when you know that a specific job needs doing, eg sheep dipping, drenching or foot treatment, set a day to do this job and stick to it no matter what, because these jobs cannot be put off.

Write out your rota for the washing up etc and pin it up on the wall so that everyone can see it. Make a list of those jobs given to the children and put their names beside the appointed task. It all saves a lot of argument in the long run.

Meals often present problems, both in terms of what to provide and making sure all the family is present when they are ready.

You may like to try this suggestion; it will take a little time and thought originally, but in the long term should save you time. Sit down and write a comprehensive batch of menus. If you do enough for two to three weeks, you will be able to re-shuffle them on a regular basis to ensure variety. Choose your main courses and accompanying vegetables; also the sweets. You can then pin up the menus for the week, so everyone can see what needs to be prepared for the main meal each day. This provides adequate information for any member of the family, who can then prepare the vegetables etc for the meal. You will also find that having menus helps you when it comes to writing your shopping list.

Having decided how frequently you need to shop, draw up your shopping list accordingly. Make sure that you buy enough to last until the next time you shop. Buying in bulk can often be cheaper anyway. You can refer to your menus to make sure you buy in anything required for the meals that perhaps you wouldn't buy normally.

Have regular meal times. This way all members of the family will know when to arrive for meals, no matter where they are or what they are doing. Obviously, you will need to be flexible on occasions, but for day-to-day living it will make life easier.

Children need some sort of routine to live by to make them feel safe and secure. It also saves a lot of arguments and misunderstandings.

They will at some stage have homework and a few ground rules should be made for them to follow, such as:

- Change from school uniform into play clothes as soon as they get home.
- Do any homework necessary and prepare school bag for the following day.
- Do any chores that have been set for that day.

Once this is achieved their time is their own and they don't suddenly decide at bedtime, or even at breakfast next morning, that they haven't done their homework, or that they need to find their swimming things for that day.

Time saving devices

There are many time saving devices these days. Some are an enormous help, others often a waste of money and space.

A freezer is a must, in order to do your bulk shopping and preserve your own produce; also to enable you to batch cook, saving time and money. Moreover, many smallholdings are several miles from the nearest town; during the winter months it may not be possible to get to town because of bad weather.

An automatic washing machine is a great time saver. It enables you to do your washing and get on with other jobs at the same time.

A tumble-dryer is an advantage during wet weather, especially if you have young children, but it is not a necessity; washing can be dried in other ways.

A microwave oven is very useful for heating pies etc; cooking vegetables and stewing fruit are quicker and easier; it is also convenient for defrosting food. Once again though, not a necessity.

Decor and furniture

Decor and furniture are really a matter of choice, but the simpler they are, the easier they are to keep clean.

Bedrooms

Children's rooms always have a lot of clutter in the way of toys. Plenty of storage space is required here. A wardrobe and chest of drawers are all that is needed for clothing, but if you put a cupboard above the wardrobe this will provide space for storing games, puzzles etc. If you can build a worktop or a dressing-table top on the chest of drawers, this will provide a dual purpose dressing table cum desk. Give facilities for studying and homework and also somewhere to sit and make models, do jigsaws etc.

The decor if kept light and plain will give an impression of space and light. The plainness can be alleviated by putting posters on the walls, brightly coloured curtains and bedding.

The master bedroom can also be kept simple, with plain light colours, possibly one or two pictures on the wall. Spacious storage is needed, wardrobe/cupboard, dressing table/chest of drawers etc. Use brightly coloured or patterned curtains and bedding.

Living room

If possible make the living room a place for relaxation only, when everyone has finished for the day and is clean and tidy. A lot of dirt is brought in from outside during the day, so it is nice to keep one room free from daily activity.

Choose restful colours for this room. Try not to have too much clutter in the way of furniture. Too many fussy bits and pieces do not create a restful atmosphere.

Dining room

You may have a separate dining room or it may be a kitchen/diner. It is a matter of personal choice. A separate dining room has its advantages: you don't have the cooking smells and, if you are entertaining, your guests don't have to see the dirty pots and pans while they are eating.

Your dining furniture is according to personal choice, but don't make it too dark or heavy.

If you have a kitchen/diner, choose your dining furniture to co-ordinate with your kitchen, eg if you have a pine kitchen, choose pine furniture for your dining area.

Kitchen

Everyone has their own views on how a kitchen should be. If you have a modern home, obviously a modern kitchen is in keeping. However, if you have a large old farmhouse you may wish to choose more traditional decor in the way of units etc. Today there is a large choice of kitchen units from modern through to traditional wood.

If you are lucky enough to have a conservatory or utility room attached to your home, make good use of it. You may wish to install your freezer, washing machine or tumble drier in there. This is also an ideal place for the family to store their outdoor clothes and shoes/wellington boots. It will keep some of the dirt out of your home.

The fact is, if you can organise yourself and your family, your day-to-day living will go so much more smoothly. Obviously you will need to bend the rules or deviate from a set plan occasionally. By no means make your life too regimented, and use them as a guide-line only.

You have chosen to live a very different way of life now. Some people find it harder to adjust than others. Young children adapt very quickly and soon settle to their new schools and surroundings; however, teenagers may find it more difficult to settle down, particularly if they have been used to going out a lot.

We live in a village community and were at first surprised at how much is offered in the way of entertainment. There are youth clubs for children aged seven and over. In one club they cover a wide range of activities from swimming, football and riding to dancing, singing and crafts. They enter county competitions for quizzes, recitation and singing. And they regularly all get together for a sing-along evening.

There are village football teams for different age groups; providing equal opportunities to both boys and girls.

There are two swimming pools within 12 miles which are open to the public.

Whist drives are held at the village hall, plus various other activities.

The village school is excellent. Being a small school, with only 50–60 pupils in the whole school, ages ranging from four to eleven years, the children get far more attention and better education than in a larger school. When you go to the school, the teachers and head-teacher are far more likely to know who your child is and what has been happening in your child's life, even if they are not your child's teacher.

You will find that because of your new way of life, appetites will increase and often children who have been rather choosey in their eating habits will become healthy and hearty eaters. Because of the space available for children to run around and play in, they become strong and healthy. You may also find that your children are more relaxed because of your way of life, now that you have escaped the 'rat-race', to a certain extent anyway.

Clothing

The most important thing to remember about clothing is that it must be practical and comfortable. A smallholding is most definitely not the place for neat skirts and blouses, expensive dresses and stylish shoes. Buy good quality for such items as waterproof clothing and working boots and shoes. It doesn't pay to skimp on these as you most certainly get what you pay for.

Women

There is nothing worse than being cold during the winter months, so choose your clothes for warmth rather than style.

Nowadays thermal underwear is very trendy and a good idea. Socks are a must; wearing wellies is none too warm and when you are ploughing your way through mud you feel even colder. Depending on how cold it is you may wear a T-shirt and jumper and find it is enough. If not try a T-shirt under a brushed cotton working shirt, a nice thick jumper to top it and you should be okay. Jeans or cord trousers are the most practical things to wear around the smallholding. Skirts can be a hindrance when you are trying to work with animals and do farm work. If you decide against thermal undies, try woolly tights under your trousers or jeans to give you that extra layer against those biting winds.

Outdoor clothing is most important and must be given serious consideration. It doesn't matter where you live in the British Isles, it rains a lot. Invest in a good wax cotton jacket, expensive maybe, but invaluable. They keep out the wind and rain and most have detachable hoods. They are quite generous in their sizing, allowing for thick jumpers or body warmers, but don't skimp on the size you require.

An old anorak or jacket can be quite adequate for those intermediate days, when a wax cotton jacket feels too heavy.

Invest in a pair of waterproof trousers which will save you a lot of washing, as well as keep you dry.

Wellies are essential, no matter what time of year it is. During the winter months you will probably find yourself wearing nothing but wellies on your feet. A stout pair of walking shoes or boots is also very useful.

Summer wear is very much a personal choice; some may prefer to wear a light-weight pair of cotton trousers, topped with a T-shirt or blouse, whereas others may prefer shorts and T-shirt. Skirts have a habit of getting in the way.

A light-weight waterproof jacket is very useful for those summer showers.

Trainers can be an advantage, but they tend to make your feet hot. You may decide that light-weight canvas shoes are more suitable. Sandals are not particularly practical around the smallholding. You could find yourself with damaged feet, because of inadequate protection while working.

Men

Men are not as particular as women about what they look like while working, but they usually have very strong likes and dislikes.

Thermal underwear is a good idea and T-shirts or brushed cotton shirts (or both)under a good thick jumper will keep you warm in winter. Jeans or cords complete the attire.

Some men wear 'boiler suit' type overalls when working around the farm. These save everyday clothes and cut down on the washing. (Women may even consider investing in some.)

Outdoor clothing is much the same as for women. It must be waterproof and windproof. You will also find an old jacket useful.

Footwear must include wellies, of course, and a good pair of working shoes or boots.

Summer wear for men is similar to women's. A lighter weight pair of trousers, maybe, plus a T-shirt or shirt. This is very much an individual choice.

A light-weight waterproof is also useful.

The suggestions for everyday wear about the smallholding mean that you keep your better clothes for those evenings out (if you're lucky), a trip to town or a visit to friends or relations. However, don't make the mistake of dressing up to go to market; stick to your jeans, jacket and wellies - you will blend in better with your fellows.

Children

A few general tips for children's clothing to begin with. Don't pay a fortune for them. Children grow so fast and before you know it, the clothes don't fit any more.

Some schools have uniform. If they don't, make a simple rule from the beginning: these clothes for school, these for best and these to play in. It will save no end of arguments and tears in the long run.

Don't buy too many 'best' clothes unless, of course, you go out a lot (which is most unlikely to begin with). What you do buy, of course, depends on the age and tastes of your child.

Play clothes for boys should be jeans or cords, topped with T-shirt and jumpers, and during the summer shorts and T-shirts.

For girls, forget she is a girl when playing. Nine out of ten small girls are tomboys at heart. Jeans or dungarees, T-shirts and jumpers are the most comfortable and practical. Dresses and skirts are very impractical for playing in, especially on a smallholding.

Outdoor clothing is the same for either sex. Waterproof, windproof jackets or coats and overtrousers are absolutely essential. The overtrousers will save no end of washing. Boys particularly always seem to manage to come in soaking wet or covered in mud. An old anorak or parka is very useful for playing in.

Footwear must, of course, include wellies, and trainers are adequate for drier days.

Whatever you dress your children in, make sure they are warm and dry, but also that they are comfortable, not just physically but mentally. If they know that it doesn't matter if they get dirty while playing, it will make for happier children.

Chapter 10

Cooking, Freezing and Appliances

Cooking

Batch cooking

There are a number of reasons why batch cooking is a good idea.

- It saves on fuel.
- It makes more use of your freezer.
- It provides meals for the times when you are too busy to prepare them.
- There's always something to fall back on when those unexpected guests arrive.

Try to set aside one day a fortnight or a month to restock your freezer. How frequently you need to do this depends on the size of your freezer and your family. Setting aside a whole day for your baking is wiser than trying to fit in a couple of hours here and there. Once your oven is hot, it is far more economical to keep filling it with items to be cooked than to switch it on to do one or two dishes. Instead of making one of a particular kind of cake, make two or three and so on.

Useful items to cook in batches are cakes, biscuits, pies (savoury and sweet), pasties, bread, casseroles and soups.

Cakes are handy for tea or lunch-time extras, and particularly for unexpected guests. Biscuits always go down well with children, as do homemade sweets, and many adults will have a biscuit or two with their tea or coffee.

Family size meat pies provide an instant and satisfying meal, whereas individual pies are useful for lunch-time snacks, as are pasties. These are also excellent when you have an extra mouth to feed at lunch-time, when extra help has been taken on for a job around the smallholding, such as shearing or haymaking time. Fruit pies make a welcome sweet, and they are a useful way of using the fruit harvested on your holding.

Homemade bread and rolls are always much nicer than shop bought items and generally will freeze well.

Casseroles can be cooked in large quantities and stored in

containers of appropriate sizes to provide a ready-made meal. Soups can also be prepared in the same way.

A conversion chart for oven temperatures is given below. It may be useful to you, when using a Rayburn, gas or electric cooker. The old Rayburns use Fahrenheit, whereas most electric cookers now use centigrade. Even recipe books vary when giving oven temperatures. Some of the very old books use such statements as 'moderate' or 'cool'.

Conversion chart

Gas mark	*°C*	*°F*	
¼	110	225	
½	130	250	very cool
1	140	275	
2	150	300	cool
3	170	325	
4	180	350	moderate
5	190	375	
6	200	400	moderate, hot
7	220	425	
8	230	450	hot
9	240	475	very hot

Many of the old recipe books are excellent. *Farmhouse Fare* is a favourite of mine. Also some of the farming books have some very good recipes.

Freezing

Freezing is possibly the quickest, easiest and most convenient way of storing food for future use. Care must be taken when preparing food for freezing, particularly where meat, poultry and fish are concerned. There is a wide range of produce that can be frozen easily.

Methods of freezing

- Free flow pack (open freezing)
- Dry sugar pack
- Syrup pack
- Puréed
- Blanching

Free flow pack

The open freezing method is used for a number of items. The

process is relatively simple. Place items on plastic trays or baking trays lined with foil. Do not let items touch one another. Place trays in freezer and leave until frozen solid. Place items in bags or containers. Seal, label and return to freezer.

Dry sugar pack
Here is a method used for storing fruit, particularly soft, juicy, whole or sliced fruit. Place items in a large shallow dish. Sprinkle with sugar, leave to stand for a few minutes so that the juices begin to flow, make sure fruit is well covered, pack in containers, label and place in freezer.

Syrup pack
Once again a method used for storing fruit and good for fruit that easily discolours. The type of syrup depends on the sweetness of the fruit and on personal taste. Put prepared fruit into containers, label and cover with syrup, seal and freeze.

Puréed
This is an excellent way of freezing over-ripe or damaged vegetables and fruit. Puréed fruit is ideal for flavouring ice creams and fools. Vegetables are useful for soups, sauces and baby foods.

Blanching
Blanching vegetables helps to preserve their colour and flavour. Wash vegetables thoroughly and prepare accordingly. It is best to blanch approximately one pound of vegetables at a time. Place the prepared vegetables in a wire basket and immerse into fast boiling water, cover and when the water comes back to the boil, start timing. After the required time, drain vegetables and plunge into iced water to ensure they are completely cooled. Open freeze as described opposite, then pack into bags or containers, seal and freeze.

Packaging, labelling and recording
It is important that food is packed properly for freezing, to prevent contamination, freezer burn or dehydration. When packaging food for the freezer, there are a number of ways to do this, eg rigid containers, sheet wrapping and bags.

Rigid containers include such items as margarine containers, ice cream tubs, plastic or foil containers. These are best used for liquids, purées, soups, casseroles, sauces or juices, but are also excellent for soft fruit, herbs, biscuits etc.

Sheet wrapping includes items such as foil, polythene wrap or waxed paper. Foil is an excellent packaging, especially for irregular shaped joints of meat etc, because it is easy to mould the foil to the shape, thus expelling all the air; however, be sure to use freezer foil, not ordinary household foil, as it is not strong enough.

Polythene wrap is also useful for irregular shaped items but, unlike foil, it needs to be sealed with tape.

Waxed paper is good for separating items within their packages, eg between chops so that they can be easily removed for use.

Labelling is important when freezing food, to enable easy identification. Means of labelling include special freezer marker pens or self-adhesive labels. We found the marker pens a bit messy and favour the self-adhesive labels. These come in various colours, so if you care to colour-code your produce, it enables you to find the required item more quickly.

Whichever way you label you must be sure to detail the type of food, quantity and date of freezing.

It is useful to keep a record book of the contents of your freezer. You can buy freezer record books, but it is quite simple to make your own. Buy an exercise book and divide it into sections for different types of food. You need to record the date each item is frozen, what it is and the quantity. When it is removed from the freezer either cross it off or put down the date used.

What to freeze

There is a vast variety of items that come under the 'home produce' heading, from baking to fruit and vegetables and meat and poultry. We shall consider each section individually and give a small example of what and how to freeze each item.

Baking covers such things as bread, cakes, biscuits, pies etc. Bread can be wrapped in foil or polythene wrap and sealed or even placed in a polythene bag; be sure to label it.

Cakes can also be wrapped in foil or polythene and sealed or placed in polythene bags; however, if you are putting a cake into a bag, wrapping it in waxed or greaseproof paper first helps. Small cakes can be placed in bags and sealed with a tie.

Biscuits are often better frozen in a rigid container, such as an ice cream container, as this prevents them from being broken.

Pies could be family size, individual or pasties even. They can be made in foil trays and cooked and frozen *in situ*, ready to be removed from the freezer and reheated in their foil trays. Fruit pies could be made and frozen without cooking, so they can be removed and cooked when required. These pies could be wrapped in foil or polythene while still in their trays and then frozen.

Freezing fruit is a lovely way of enjoying such things as raspberries out of season. Often when fruit is in season there can be a lot of waste, because it is so plentiful it is impossible to use it all at once. This is where freezing comes in. Freeze all surplus fruit, but it must be in perfect condition otherwise it will not keep. Some suggestions for using damaged fruit are given below.

Apples freeze well. They can either be peeled, cored and sliced, dipped in a lemon solution to prevent discoloration, open frozen then packed in bags, or puréed and packed in rigid containers. Blackberries can be open frozen, with or without sugar then packed in rigid containers or bags. Blackcurrants can be packed into bags and frozen or puréed and packed into rigid containers. Plums can be stoned and packed in rigid containers using a medium syrup. Raspberries can be either rolled in sugar or left unsugared, open frozen and packed in bags or rigid containers. Rhubarb can be chopped into pieces, open frozen and packed into bags. Strawberries can be open frozen and packed into containers.

There are many other kinds of fruit, but the above are some of the most common that you are likely to produce yourself.

A good way of making use of large quantities of damaged fruit is to turn it into wine. Fruit wines can be very pleasant and are relatively cheap to produce. Some varieties of damaged vegetables also make good wine.

Small quantities of damaged fruit need to be used quite quickly, because they decay rapidly. The damaged parts can be removed and the remainder used in trifles, fresh fruit salads and fruit juices for immediate use.

Bulk buys

It is cheaper to buy in bulk when stocking your freezer: items such as meat, vegetables (if home grown are not available), ice cream, beefburgers, fishfingers etc. Meat can be bought from an abattoir at very good prices, but you must remember that here you have to buy, say, a pig or a whole lamb to get the low price.

When buying frozen vegetables, it is more sensible to buy big bags than numerous smaller bags; however, don't be lulled into thinking that buying the cheapest brand is the best saving as often this is not the case. Buy big boxes of fishfingers and beefburgers, it is much cheaper, but be sure to choose a brand that you know and like. You can then be sure that there will be no waste.

What convenience foods you buy for your freezer is a matter of personal choice.

Other methods of storing

Bottling

This method of preserving involves sterilizing your fruit. This can be done either by the water bath method or a pressure cooker. Proper preserving jars must be used as they must stand up to the heat of the water when processing.

The jars must be thoroughly clean and warm before packing fruit into them. Pack fruit in tightly, but be careful not to squash it. Use only perfect fruit, to prevent spoiling.

Pour syrup over the fruit, dispersing any air bubbles by inserting the blade of a knife down the side of the jar.

Place seals on jars, either glass or metal, and secure with clips or screw bands. If you use screw bands, screw them down and then loosen by a quarter of a turn to prevent the jars exploding during processing.

When using the water bath method, place the jars on a trivet in a large pan and pour in water to the level of the syrup. Heat the water gradually so that it reaches 130°F in one hour, and the required temperature after 1½ hours. The final temperature varies according to what you are bottling. This should then be held for 10–30 minutes depending on the type of fruit. After processing, remove the jars and tighten screw bands immediately. Leave to stand overnight.

When using a pressure cooker ensure that it is deep enough to stand the jars in and that the pressure can be maintained at 5lb. Put one inch of water in the pan, place the jars on the trivet, cover the pan, but leave the vent open until the steam flows. Close the vent and bring up to pressure; hold for 1–5 minutes depending on the type of fruit. Leave the pan to cool for 10 minutes before removing the lid. Take out the jars and tighten the screw bands as before. Leave overnight.

Pickling

This is preserving fruit or vegetables in vinegar. Many varieties of vegetables can be pickled, but the most common are onions, beetroot, cucumber and red cabbage. White or malt vinegar can be used. Both have their advantages. White vinegar gives a better colour and malt vinegar a better flavour, but both can be flavoured with spices.

Large vegetables can be cut into small pieces or shredded (such as red cabbage). Small vegetables, such as onions and beetroot, can be pickled whole. This is an ideal opportunity to use those small beetroot not used in the summer salads; they are ideal for pickling.

Vegetables must be washed thoroughly and placed in a bowl and generously sprinkled with salt in layers. This will draw out excess water. Leave to stand overnight. Next day wash the vegetables and drain them before packing into clean jars. Pour in spiced vinegar to cover the vegetables to half an inch above. Seal the jars with vinegar-proof tops. Store for approximately one month before use.

When doing sweet pickles, ie fruit, leave out the peppercorns when spicing the vinegar. Instead use spiced vinegar as a base for a syrup to pickle fruit in, allowing 2lb of sugar to each pint of vinegar. Otherwise continue as for ordinary pickles.

Jams, jellies and marmalades

Here, in all three cases, good quality fruit must be used. It must also not be over-ripe, as it will not set properly.

A jam is made when you cook the fruit with sugar.

A jelly is made from cooking the fruit first and straining off the juices and using the juice to cook with the sugar to produce the jelly.

A marmalade is made of citrus fruit only and usually the peel is also used.

Drying

Today the drying method of preserving is generally used for herbs, although it is one of the earliest forms of preserving food. In the past items such as cereals, berries, fruit and meat were dried.

The object of drying is to remove all moisture to prevent enzymes causing deterioration.

When drying herbs at home the two main points to remember are: (a) use the correct temperature and (b) plenty of ventilation. The ideal temperature is between 50°C and 65°C.

The best times for harvesting your herbs vary. If using the leaf, harvest when the flowers are in bud. If using the flower, harvest just before it is fully open. Harvest seeds when the heads turn brown.

Another point to remember is, when using the leaves or flowers, harvest in the early morning but not until the dew is gone.

How to dry
Air drying is the easiest method. Tie sprays of herbs into bunches and hang them in an airy room, loft or garage, but be careful of petrol fumes. They should be ready after two or three weeks.

Quick drying keeps a better aroma. Spread herbs on to newspaper, cotton or muslin which has been stretched over a cake tray. Put into a warm, dry place. Leave for approximately 12 hours.

Storing. To store, strip leaves from stems and crumble coarsely and put them into clean, air-tight containers. Dried flowers and seeds can be placed straight into a container.

Vegetables
Potatoes can be lifted and stored in paper sacks and kept in a cool, dark place.

Parsnips, swedes and turnips can be left in the ground and dug as required.

Carrots and beetroot can be stored indoors between layers of sand or peat.

Onions can be stored in net bags or strung.

Appliances

Heating
There are numerous ways of heating your home. The best method depends on the size of your home and the availability of fuels. A few of the advantages and disadvantages of each type are given below.

Natural gas is a very efficient form of heating. It can not only heat your home, but also your water. Boilers can be wall-mounted, saving space, and radiators placed in each room. It is a clean method of heating and can be switched on or off automatically or by the flick of a switch. You could also have a gas fire, say, in your living room if required. Radiators can usually be individually regulated allowing you to have a different temperature in each room.

However, gas is not available in all areas. Towns and a few villages may have gas, but the more outlying areas do not.

Calor gas is a clean type of heating and is similar to natural gas in its uses and is generally used as the alternative to it.

The main difference between the two is that natural gas is piped whereas Calor gas is run either from a bottle or tank. At the time of writing, the cost of a bottle is approximately £11.35 a quarter. The installation of a tank costs approximately £335 and the standing charge is £12 a quarter; the gas, however, is cheaper by the tanker than the bottle.

Boilers vary from £300 to £800, the wide variation in prices being accounted for by the facilities the various models offer. So it depends what you require of your boiler as to the price you will have to pay.

Fires with backboilers vary from £490 to £740. These prices cover the difference in fire fronts. You can have just a Calor gas fire, without a backboiler; they range from £95 to £350 depending on fire front. As a back-up to Calor central heating they work very well. If you have a fire only it is obviously not as efficient and is a more expensive way of heating.

Oil-fired central heating can work out cheaper than bottled gas.

Electricity is a clean and easy form of heating. There are a number of different forms of electric heating, such as Total Heating, Economy 7 and Dimplex storage heaters. There are many different types of heaters and storage heaters, far too many to cover here, and choice is purely personal. Electric fires can be used as focal points in living rooms etc. As with the Calor gas fires, these depend on what frontage you choose.

Solid fuel heating covers a fairly wide range, from a coal or log fire to a sophisticated solid fuel heater, ie Rayburn/Aga.

A coal or log fire merely heats one room and is not very efficient. They are a very attractive feature in a lounge or dining room and can be useful on cool autumn evenings, but you must remember that they need cleaning out and lighting each day. Chimneys have to be swept regularly and you have to bring in coal or logs to feed them.

If you have a plentiful supply of logs a wood-burning stove may be the answer, but again these need to be stoked up regularly.

The ideal use of this form of heating is to install a dual purpose heater/cooker, ie Rayburn. There are many different types of these heaters, some very simple, others very sophisticated. Their prices range from approximately £100 (second-hand) to £2000. They can be fitted with backboilers to heat your water and run a central heating system. The size of the boiler required depends on the amount of hot water and the size of your home.

They need coal and/or wood; they also need cleaning out and

chimneys need cleaning, but they are the most versatile form of solid fuel heating for the smallholder.

To sum up on heating, remember your situation before you decide upon your form of heating. Calor is very efficient, clean and easy, as is electricity, whereas solid fuel is not so easy or clean. But don't rely completely on electricity: smallholdings are often situated in remote areas where electrical failure, especially during winter months, is quite common. Have some form of solid fuel heating as an alternative.

We successfully run a solid fuel Rayburn with a high output boiler fitted, which runs our central heating and heats our water. We also run an electric cooker alongside this. During the winter a Rayburn heats our water, runs our heating and we can use it for cooking as well. During the summer we use the electric cooker for cooking and an immersion heater for our water.

You could use another combination, eg Rayburn (or other make) and a Calor gas cooker and immersion heater.

Cookers

If you are used to natural gas for cooking, your nearest choice would be Calor gas. Free standing cookers range from £250 to £740. Built-in hobs and ovens range from £569 to £820. They all vary in what they offer, as do all cookers. If you use Calor only for cooking, you would only need a bottle to run it from. These cost less to run per quarter than electric cookers.

Free-standing, electric cookers range from £250 to £650 plus; the built-in hobs and ovens are obviously more expensive.

Solid fuel cookers have been covered in the heating section (see page 121).

The advantages of the Calor gas cooker is that you can use it all year round, it is not affected by power cuts, and it is clean and easy.

The advantages of the electric cooker are much the same as for Calor gas; the disadvantage is the possibility of power cuts.

Solid fuel cookers are great in winter, as they also keep your kitchen warm, but they make the kitchen excessively hot in summer and are not so clean.

Washing machines

There are two main types of washing machine, automatic and twin-tub. Both are very good, but the automatic is more sensible for a family who have other commitments. You can put your washing in, put the powder in and set the programme, then let it get on with it, while you do something else.

With a twin-tub you need to be in constant attendance, to move the washing from the machine to the spinner section. It involves more effort and time.

Twin-tubs are generally cheaper than automatics, the prices depending on the area and the shop you buy your machine from; go to local auctions if you are buying second-hand.

A point to watch for is that if your water supply is spring fed you may not have the pressure required to operate an automatic machine, so check your water pressure and check with the manufacturer's instructions what pressure is required by the machine.

A tumble drier is a very useful appliance, but not essential. It saves having wet washing lying around and is relatively cheap to run. It can be particularly useful if you have young children or a baby, when there always seems to be a lot of washing.

Tumble driers come with several different methods of venting. Some have a back vent that needs a hole made in the outside wall so that all the steam goes outside. Some have hose-type attachments; these have to be hung out of a window or door to get rid of the steam. Others have a vent in the door which sends the steam directly into the room where it is situated. The final sort is fitted with a condenser; this needs no vent at all.

Washer/driers are fitted with condensers and are useful if you are short of space. They are more expensive than an automatic washing machine, but if you consider the price of an automatic machine and a separate drier added together a washer/drier is a better deal.

Fridges and freezers

Most fridges are electric, but you can buy a Calor gas fridge. A Calor gas fridge will cost you approximately £600 to buy and in the region of £9.95 a month to run.

The electric fridge is more widely available in a variety of styles and sizes. There are larder fridges, fridges with small ice boxes, large fridges with fair sized freezer compartments, or fridge/freezers, where the size of the freezer varies from small to the same size as the fridge itself.

Prices vary greatly, depending on the type of fridge or fridge/freezer you choose and where you buy it. A small fridge starts at around £129 and rises according to facilities included. Fridge/freezers start at about £200 and prices increase with size.

The fridge is a most useful appliance and is almost definitely a must. It enables you to store such items as milk, bacon, cheese, butter etc, all items that can soon go bad if not kept cool. It also enables you to keep meat fresh for a short period. A larder fridge gives you plenty of room for your goods, but it does not have the advantage of a freezer compartment.

A fridge with a freezer compartment enables you to keep frozen goods for a period. Freezer compartments are star rated. One black star means that frozen food can be stored for one week and ice cream for one day. Two stars mean that frozen food can be kept for one month and ice cream for two weeks. Three black stars mean that frozen food can be kept for three months and ice cream for one month.

A white star in front of three black stars on a freezer, denotes that it is suitable for freezing fresh food. Fresh food can only be frozen properly in a freezer.

Freezers are a must for people intending to live a distance from shops. A freezer enables you to bulk buy and batch cook. It provides convenient storage for meat, bread etc (see page 114). There are two types of freezer, the chest and the upright.

Chest freezer. The size starts at about 4 cubic feet and rises to approximately 20 cubic feet. It takes up more floor space than an upright, therefore siting it may cause a problem. It can be sited in an out-building, garage, under the stairs or in a basement.

The running costs of a chest freezer are more economical than for an upright freezer, as when the lid is opened less cold air escapes.

You must pack your chest freezer sensibly, remembering that you will have to remove items on the top to reach those on the bottom. You can buy wire partitions to section your freezer, enabling you to divide your types of food accordingly. You can also buy wire baskets to hang on the top ledge of the freezer. These are useful for storing small items.

Upright freezer. This is a better buy if you are limited for floor space, but it is not as economical as a chest freezer to run. Depending on the style of the freezer, you may have adjustable shelves throughout or there may be a combination of shelves and pull-out drawers. Some models have sectioned doors to enable storage of smaller items.

Shop around for your freezer. Prices vary greatly but the cheapest is not always the best; bear in mind the back-up services for maintenance etc.

Which type of freezer you choose is purely personal. Take into consideration the size of your family, where you are going to site your freezer, what you intend to store in it, ie home produce, convenience foods or a combination of both and, most of all, how much you can afford to spend.

Although we have said that some items are essential and others are not, it does, of course, depend on the individual as to what is classed as 'luxury' or 'essential'; obviously circumstances vary from family to family and what we consider a luxury, you may consider an essential.

Chapter 11

Useful Addresses

Production and distribution

British Wool Marketing Board
Oak Mills, Station Road, Clayton, Bradford, West Yorkshire BD14 6JD

Commercial Rabbit Association
Little Onn House, Little Onn, Stafford ST20 0AU

Crayfish farming
British Crayfish Marketing Association Ltd
Riversdale Farm, Stour Provost, Gillingham, Dorset

Meat production
Meat and Livestock Commission
PO Box 44, Queensway House, Bletchley MK2 2EF

Pigs
National Pig Breeders' Association
7 Rickmansworth Road, Watford, Hertfordshire WD1 7HE

Quail farming
Reiver Game
87 Dene Road, Wylam, Northumberland NE41 8HB

Worm farming
Wonder Worms
Pine Trees Farm, Hubberton, Sowerby Bridge, West Yorkshire HX6 1NT

Distributor of organically grown vegetables
Organic Farm Foods
Llanbed Industrial Estate, Lampeter, Dyfed, Wales

Distributor of organically produced meat
The Real Meat Company
Easthill Farm, Warminster, Wiltshire

Information on all rare breeds and source of the magazine, *The Ark*

Rare Breeds Survival Trust
4th Street, National Agricultural Centre, Kenilworth, Warwickshire CV8 2LG

Booklets on various aspects of farming, designed to be informative while advertising ICI's products.
ICI Agricultural Division
Farm Advisory Service, PO Box 1, Billingham, Cleveland

Small Firms Division

Department of Employment
Steel House, Tothill Street, London SW1H 9NF
The Department has established a number of regionally based Small Firms Centres: telephone the operator on 100 and ask for freefone Enterprise to be put through to your nearest Centre. The same service is also offered by:

Northern Ireland
Local Enterprise Development Unit, Ledu House, Upper Galwally, Belfast BT8 4TB; 0232 491031
Northern Ireland Development Agency
Maryfield, 100 Belfast Road, Holywood, County Down; 02317 4232
Scotland
21 Bothwell Street, Glasgow G2 7NR; 041-248 6014
The Scottish Development Agency
Rosebery House, Haymarket Terrace, Edinburgh EH12 5EZ; 031-337 9595
Highlands and Islands Development Board
Bridge House, Bank Street, Inverness IV1 1QR;0463 234171
Wales
16 St David's House, Wood Street, Cardiff CF1 1ER; 0222 396116
The Welsh Development Agency
(Small Business Division), Treforest Industrial Estate, Pontypridd, Mid Glamorgan CF37 5UT; 0443 852666
Agricultural Development Advisory Service *(advice for farmers)*
Ministry of Agriculture, Fisheries and Food, Great Westminster House, Horseferry Road, London SW1P 2AB; 01-216 6311

Agricultural Mortgage Corporation Ltd
Bucklersbury House, 3 Queen Victoria Street, London EC4N 8DU; 01-236 5252

Alliance of Small Firms & Self Employed People
33 The Green, Calne, Wiltshire SN11 8DJ; 0249 817003

British Insurance Association
Aldermary House, Queen Street, London EC4N 1TT; 01-248 4477

Business in the Community
227A City Road, London EC1V 1LX; 01-253 3716
and
Romano House, 43 Station Road, Edinburgh EH12 7AF; 031-334 9876

Equipment Leasing Association
18 Upper Grosvenor Street, London W1X 9PB; 01-491 2783

Federation of Agricultural Cooperatives (UK) Ltd
17 Waterloo Place, Leamington Spa, Warwickshire CV32 5LA; 0926 450445

Law Society
Legal Aid Department, 113 Chancery Lane, London WC2A 1PL; 01-242 1222

Mail Order Traders' Association
25 Castle Street, Liverpool L2 4TD; 051-236 7581

Market Research Society
175 Oxford Street, London W1R 1TA; 01-439 2585

National Association of Shopkeepers
Lynch House, 91 Mansfield Road, Nottingham NG1 3FN; 0602 475046

National Farmers' Union
Agriculture House, Knightsbridge, London SW1X 7LY; 01-235 5077

National Federation of Self-employed and Small Businesses Ltd
32 St Anne's Road West, Lytham St Annes, Lancashire FY8 1NY; 0253 720911
and
140 Lower Marsh, London SE1 7AE; 01-928 9272

Office of Fair Trading
Field House, Bream's Buildings, London EC4A 1PR; 01-242 2858

Rural Development Commission
(incorporating COSIRA), 11 Cowley Street, London SW1P 3NA; 01-222 9134

Chapter 12

Further Reading

Production and distribution

Backyard Pig and Sheep Book Ann Williams (Prism Press)
Farm Management Pocketbook John Nix (Wye College)
Profitable Sheep Farming M Cooper and R J Thomas (Farming Press Ltd)
TV Vet Sheep Book (Farming Press Ltd) (There are similar books available in this series for other types of stock.)

Cooking, preserving and freezing

Farmhouse Fare (Hamlyn)
Farmhouse Kitchen 1, 2 and 3 (Yorkshire Television Enterprises Ltd)
Home Preserving (Marshall Cavendish Ltd)
The St Michael All Colour Freezer Cookery Book, Madeline Fraser (Marks and Spencer Ltd)

Business books from Kogan Page

Business Rip-Offs and How to Avoid Them, Tony Attwood
Buying for Business Tony Attwood
Debt Collection Made Easy, Peter Buckland
Export for the Small Business, 2nd edition, Henry Deschampsneufs
Financial Management for the Small Business, 2nd edition, Colin Barrow
Getting Sales, Richard D Smith and Ginger Dick
The Guardian Guide to Running a Small Business 7th edition, ed Clive Woodcock
How to Buy a Business, 2nd edition, Peter Farrell
How to Cut Your Business Costs, Peter D Brunt
How to Prepare a Business Plan, Edward Blackwell
Law for the Small Business, 6th edition, Patricia Clayton
Raising Finance: The Guardian Guide for the Small Business, 3rd edition, Clive Woodcock
Running Your Own Boarding Kennels, Sheila Zabawa

Running Your Own Mail Order Business, Malcolm Breckman
The Stoy Hayward Business Tax Guide (annual)
Successful Expansion for the Small Business, M J Morris
Successful Marketing for the Small Business, 2nd edition, Dave Patten
Which Business? How to Select the Right Opportunity for Starting Up, Stephen Halliday

Index